Solutions and Tests For Exploring Creation With Biology

Manufactured in the United States of America
Sixth Printing 2003

Published By
Apologia Educational Ministries, Inc.

Printed by
The C.J. Krehbiel Company
Cincinnati, OH

Exploring Creation With Biology

Solutions and Tests

TABLE OF CONTENTS

Solutions to the Study Guide

Tests

Answers to the Tests

TEACHER'S NOTES

Exploring Creation With Biology

Dr. Jay L. Wile and Marilyn Durnell

Thank you for purchasing *Exploring Creation With Biology.* We designed this modular course specifically to meet the needs of the homeschooling parent. We are very sensitive to the fact that most homeschooling parents do not know biology very well, if at all. As a result, they consider it nearly impossible to teach to their children. This course has several features that make it ideal for such a parent.

1. The course is written in a conversational style. Unlike many authors, we do not get wrapped up in the desire to write formally. As a result, the text is easy to read and the student feels more like he or she is *learning*, not just reading.

2. The course is completely self-contained. Each module includes the text of the lesson, experiments to perform, questions to answer, and a test to take. The solutions to the questions are fully explained, and the test answers are provided. The experiments are written in a careful, step-by-step manner that tells the student not only what he or she should be doing, but also what he or she should be observing.

3. Most importantly, this course is Christ-centered. In every way possible, we try to make the science of biology glorify God. One of the most important things that you and your student should get out of this course is a deeper appreciation for the wonder of God's Creation!

We hope that you enjoy using this course as much as we enjoyed writing it!

Pedagogy of the Text

(1) There are two types of exercises that the student is expected to complete: "on your own" questions, and an end-of-the module study guide.

- The "on your own" questions should be answered as the student reads the text. The act of answering these questions will cement in the student's mind the concepts he or she is trying to learn. Answers to these problems are in the student text

- The study guide should be completed in its entirety after the student has finished each module. Answers to the study guide questions are in this book.

The student should be allowed to study the solutions to the "on your own" questions while he or she is working on them. When the student reaches the study guide, however, the solutions should be used only to check the student's completed work.

(2) In addition to the solutions to the study guides, there is a test for each module in this book, along with the answers to the test. **We strongly recommend that you administer**

each test once the student has completed the module and all associated exercises. The student should be allowed to have only pencil, paper, and any tables that are specifically mentioned in the test. We understand that many homeschoolers do not like the idea of administering tests. However, if your student is planning to attend college, it is *absolutely* necessary that he or she become comfortable with taking tests!

(3) All definitions presented in the text are centered. The words will appear in the study guide and their definitions need to be memorized.

(4) Words that appear in bold-face type in the text are important terms that the student should know.

(5) The study guide gives your student a good feel for what we require him or her to know for the test. Any information needed to answer the study guide questions is information that the student must know for the test. Sometimes, tables and other reference material will be provided on a test so that the student need not memorize it. You will be able to tell if this is the case because the questions in the study guide which refer to this information will specifically tell the student that he or she can use the reference material.

Experiments

The experiments in this course are designed to be done as the student is reading the text. We recommend that your student keep a notebook of these experiments. This notebook serves two purposes. First, as the student writes about the experiment in the notebook, he or she will be forced to think through all of the concepts that were explored in the experiment. This will help the student cement them into his or her mind. Second, certain colleges might actually ask for some evidence that your student did, indeed, have a laboratory component to his or her biology course. The notebook will not only provide such evidence but will also show the college administrator the quality of the biology instruction that you provided to your student. We recommend that you perform your experiments in the following way:

- When your student gets to the experiment during the reading, have him or her read through the experiment in its entirety. This will allow the student to gain a quick understanding of what her or she is to do.

- Once the student has read the experiment, he or she should then start a new page in his or her laboratory notebook. The first page should be used to write down all of the data taken during the experiments and perform any exercises discussed in the experiment.

- When the student has finished the experiment, he or she should write a brief report in his or her notebook, right after the page where the data and exercises were written. The report should be a brief discussion of what was done and what was learned.

- **PLEASE OBSERVE COMMON SENSE SAFETY PRECAUTIONS. The experiments are no more dangerous than most normal, household activity. Remember, however, that the vast majority of accidents do happen in the home!**

- Although many of the experiments in the course require a microscope, there are aspects to some of those experiments that do not. If you do not have a microscope, your student should still read through the microscope labs and perform all non-microscope exercises that the lab discusses. For example, experiment 2.1 is a trip to a pond. Although the main reason for the trip is to gather water samples for later microscope experiments, the experiment does ask the student to observe the habitat and make some drawings. Even if you do not have a microscope, the student should still visit a pond, examine the habitat, and make the drawings. The student need not gather the water, however.

Question/Answer Service

For all those who use my curriculum, I offer a question/answer service. If there is anything in the modules that you do not understand - from an esoteric concept to a solution for one of the problems - just write down your question and send it to me by any of the means listed on the **NEED HELP?** sheet that came with this course

A Word About Grading

The sciences are, by far, the most difficult of all subjects to study. As a result, students often perform significantly worse in courses like biology, chemistry and physics than they do in all other subjects, including math. This often makes the student feel that he or she is not talented in the sciences, because that's where the student gets his or her lowest grades. Often, however, this is not the case. Some of the best scientists I know received lower grades in the sciences than in any of their other courses. In fact, my own *lowest* GPA in college was my chemistry GPA. Thus, just because a straight-A student gets B's or C's in this course, they should not be discouraged from taking more science courses.

Public schools have long recognized this fact, so they implement strategies that tend to "boost" their students' grades in the sciences. For example, all public schools give their students grades on their labs and homework. Since labs and homework are always performed with the help of the teacher and fellow students, the grades on these assignments are usually quite high. This tends to boost the lower test scores, allowing students to have grades that are comparable to their other courses. Since all public schools do this, and since college admissions people (or job interviewers) will be comparing your student to publicly-schooled students, you should probably do the same. Here are my suggestions on how to grade your students:

1. Give them a grade for each lab that they do. This grade should not reflect the accuracy of the student's results. Rather, it should reflect how well the student followed directions and how well he or she wrote up the lab in his or her lab notebook.

2. Give them a grade for each test. If a test problem contains multiple parts, it should be worth more points than other test questions that do not. As a general rule, I would say that every answer that a student must write down is worth one point. That way, their percentage grade can be calculated as total number of correct answers divided by the total number of answers given. Additionally, you can give partial credit. If a student plowed through the entire problem correctly but just messed up on the calculator, the student should receive 3/4 of a point. If the student got the first couple of steps correct and messed up after that, they should receive 1/2 of a point. Of course, this grading technique requires that you learn the subject right along with the student. This is, of course, what I recommend that you do to begin with!

3. The student's overall grade in the course should be weighted as follows: 35% lab grade, 65% test grade. A straight 90/80/70/60 scale (100%-90% is an A, 89%-80% is a B, etc.) should be used to calculate the student's letter grade. This is typical for most public schools.

Finally, I must tell you that I pride myself on the fact that this course is user-friendly and reasonably understandable. At the same time, however, *it is not EASY*. This is a tough course. I have designed it so that any student who gets a "C" or better on the tests will be VERY well prepared for college.

Solutions To The

Study Guides

SOLUTIONS TO THE STUDY GUIDE FOR MODULE #1

1. a. Metabolism - The process by which a living organism takes energy from its surroundings and uses it to sustain itself, develop, and grow

b. Photosynthesis - The process by which a plant uses the energy of sunlight and certain chemicals to produce its own food. Oxygen is often a by-product of photosynthesis.

c. Herbivores - Organisms that eat plants exclusively

d. Carnivores - Organisms that eat only organisms other than plants

e. Omnivores - Organisms that eat both plants and other organisms

f. Producers - Organisms that produce their own food

g. Consumers - Organisms that eat living producers and/or other consumers for food

h. Decomposer - Organisms that break down the dead remains of other organisms

i. Autotrophs - Organisms that are able to make their own food

j. Heterotrophs - Organisms that depend on other organisms for food

k. Receptors - Special structures or chemicals that allow living organisms to sense the conditions of their surroundings

l. Asexual reproduction - Reproduction accomplished by a single organism

m. Sexual reproduction - Reproduction that requires two organisms, a male and a female

n. Inheritance - The process by which physical and biological characteristics are transmitted from the parent (or parents) to the offspring

o. Mutation - An abrupt and marked difference between offspring and parent

p. Hypothesis - An educated guess that attempts to explain an observation or answer a question

q. Theory - A hypothesis that has been tested with a significant amount of data

r. Scientific Law - A theory that has been tested by and is consistent with generations of data

s. Microorganism - A living creature that is too small to see with the naked eye

t. Abiogenesis - The theory that, long ago, very simple life forms spontaneously appeared through random chemical reactions

u. Prokaryotic cell - A cell that has no distinct, membrane-bound organelles

v. Eukaryotic cell - A cell with distinct, membrane-bound organelles

w. Species - A unit of one or more populations of individuals that can reproduce under normal conditions, produce fertile offspring, and are reproductively isolated from other such units

x. Binomial nomenclature - Naming an organism with its genus and species name

y. Taxonomy - The science of classifying organisms

2.

- All life forms contain deoxyribonucleic acid (DNA).
- All life forms have a method by which they extract energy from the surroundings and convert it into energy that sustains them.
- All life forms can sense changes in their surroundings and respond to those changes.
- All life forms reproduce.

3. Carnivores eat non-plants. This means they depend on other organisms for food, making them heterotrophs, which are also known as consumers.

4. If the tentacles are cut off, then the organism has no receptors, which sense the conditions of the environment. Thus, sensing change in the surroundings and responding to those changes will be impossible for this wounded creature.

5. These organisms reproduce sexually. In sexual reproduction, the offspring's traits are a blend of the parents, their parents, and so on. This would account for the differences between parent and offspring.

6. Science cannot prove anything.

7. In the scientific method, a person starts by making observations. The person then develops a hypothesis to explain those observations or to answer a question. The person (often with the help of others) then designs experiments to test the hypothesis. After the hypothesis has been tested by a significant amount of data and is consistent with all of it, then it becomes theory. After more testing with generations of data, the theory could become a scientific law.

8. The story of spontaneous generation shows how almost 2,000 years of executing the scientific method resulted in a law that was clearly wrong. Thus, you can't put too much faith in scientific laws. They are fallible.

9. The wise person trusts the Bible, because it is infallible.

10. Abiogenesis is a theory that states that life sprang from non-living chemicals eons ago. This is an example of spontaneous generation, a former law that said life could arise from non-life. We now know that this law is wrong.

11. Kingdom, Phylum, Class, Order, Family, Genus, Species

12. Animalia - Since it is multicelled, it is not Monera or Protista. In addition, it is not Plantae because it is not an autotroph (consumers are heterotrophs) and it is not Fungi because it is not a decomposer.

13. Monera - Prokaryotic cells belong to this kingdom.

14.
1. macroscopic, proceed to key 3
3. heterotrophic, proceed to key 5
5. decomposer, kingdom Fungi

Kingdom: Fungi (This is as far as we can go using this key.)

SOLUTIONS TO THE STUDY GUIDE FOR MODULE #2

1. a. Pathogen - An organism that causes disease

   b. Saprophyte - An organism that feeds on dead matter

   c. Parasite - An organism that feeds on a living host

   d. Respiration - The process by which food is converted into useable energy for life functions

   e. Aerobic respiration - Respiration that requires oxygen

   f. Anaerobic respiration - Respiration that does not require oxygen

   g. Steady state - A state in which members of a population die as quickly as new members are born

   h. Conjugation - A temporary union of two organisms for the purpose of DNA transfer

   i. Plasmid - A small, circular section of extra DNA that confers one or more traits to a bacterium

   j. Transformation - The transfer of a "naked" DNA segment from a nonfunctional donor cell to that of a functional recipient cell

   k. Endospore - The DNA of a bacterium that is coated with several hard layers

   l. Strains - Organisms from the same species that have markedly different traits

2. *See Figure 2.1 for the answers.*

3. a. Pilus - To grasp onto surfaces. It can also be used to grasp onto another bacterium during conjugation.

b. Capsule - To adhere to surfaces as well as to ward off infection-fighting agents

c. Cell Wall - To keep the interior of the cell together and to hold the cell's shape

d. Plasma Membrane - To negotiate what materials pass into and out of the cell as well as to metabolize nutrients

e. Cytoplasm - To hold the DNA and ribosomes in place

f. Ribosomes - To make proteins

g. DNA - To store the information needed to make this organism a living organism

h. Flagellum - To move the bacterium from place to place.

4. Most bacteria are heterotrophic decomposers.

5. Parasites, by definition feed on something produced by a host. They therefore cannot make their own food. This makes them heterotrophic.

6.
a) Asexual reproduction - By far the most common. A single bacterium copies its DNA and then divides in two. Both bacteria can then reproduce over and over again.

b) Sexual reproduction - Also called conjugation. In this form of reproduction, a donor and recipient form a temporary bridge across which a DNA plasmid from the donor travels into the recipient. This passes a trait or traits from the donor to the recipient.

c) Transformation - In this process, a bacterium absorbs the DNA of a dead bacterium, taking on the trait specified by the DNA absorbed.

d) Endospore formation - In this process, the DNA is protected by a hard shell to survive extreme conditions. If conditions return to normal, the cell bursts out of the endospore and returns to normal functions.

7. Before the steady state, the bacteria are reproducing essentially unchecked. Thus, the population increases. After the steady state, bacteria begin to die due to lack of resources. Thus, the population decreases.

8. Sexual reproduction can pass a trait from one bacterium to another. If that trait allows the recipient to survive conditions that it otherwise wouldn't, then the population is affected, because the recipient continues to live and reproduce asexually.

9. Coccus - Spherical
Bacillus - Rod-shaped
Spirillum - Helical

10. If it is Gram-negative, it is in phylum Gracilicutes. In this phylum, two classes contain photosynthetic bacteria, which are autotrophs. Since this bacterium is heterotrophic, it must belong to the only other class, class Scotobacteria.

11. Gram-positive means phylum Firmicutes. Since it is spirillum-shaped, it is neither coccus nor bacillus. Thus, it is in class Thallobacteria.

12. Bacteria with no cell walls belong to the phylum Tenericutes, which has only one class, class Mollicutes.

13. To grow and reproduce, most bacteria need: Moisture, moderate temperatures, nutrition, and darkness.

14. To reduce the chance of bacterial infection you can:

a) Heat the food so that most bacteria die and then seal it away from fresh air.
b) Dehydrate the food.
c) Freeze the food.
d) Pasteurize it.
e) Keep it in the refrigerator.

SOLUTIONS TO THE STUDY GUIDE FOR MODULE #3

1. a. Pseudopod - A temporary, foot-like extension of a cell, used for locomotion or engulfing food

b. Nucleus -The region of a eukaryotic cell which contains the DNA

c. Vacuole - a membrane bound "sac" within a cell

d. Ectoplasm - The thin, watery cytoplasm near the plasma membrane of some cells

e. Endoplasm - The dense cytoplasm found in the interior of many cells

f. Flagellate - A protozoan that propels itself with a flagellum

g. Pellicle - A firm, flexible coating outside the plasma membrane

h. Chloroplast - An organelle containing chlorophyll for photosynthesis

i. Chlorophyll - A pigment necessary for photosynthesis

j. Eyespot - A light-sensitive region in certain protozoa

k. Symbiosis - Two or more organisms of different species living together so that each benefits from the other

l. Cilia - Numerous short extensions of the plasma membrane used for locomotion

m. Spore - A reproductive cell with a hard, protective coating

n. Plankton - Tiny organisms that float in the water

o. Zooplankton - Tiny floating organisms that are either small animals or protozoa

p. Phytoplankton - Tiny floating photosynthetic organisms, primarily algae

q. Thallus - The body of a plant-like organism that is not divided into leaves, roots, or stems

r. Cellulose - A substance made of sugars. It is common in the cell walls of many organisms.

s. Holdfast - A special structure used by an organism to anchor itself

t. Sessile Colony - A colony that uses holdfasts to anchor itself to an object

2. There is no real answer for this question. Just be sure that you can name the subkingdom and phylum of each organism in Figure 3.1 when you see its picture.

3. *Euglena* and *Spirogyra*. Each of these organisms use chlorophyll for photosynthesis and thus have chloroplasts. The other two genera contain exclusively heterotrophic organisms, which obviously do not use photosynthesis.

4. A contractile vacuole collects excess water in a cell and pumps it out to reduce the pressure inside the cell. This keeps the cell from exploding. The food vacuole, on the other hand, stores food while it is being digested and has nothing to do with excess water or pressure.

5. Endoplasm is thick while ectoplasm is thin and watery. Endoplasm is found in the central region of the cell while ectoplasm is found near the plasma membrane.

6. The amoeba uses pseudopods which it creates by deforming its body. The euglena, on the other hand uses a flagellum. There is one bit of similarity. When it wants to move quickly, the euglena deforms its body in an almost earthworm-type motion. This is used to enhance the motion supplied by the flagellum, and is something like the amoeba's motion.

7. There are more than three, but you only need to recall three. The ones we discussed in this module are *Entamoeba histolytica, Trypanosoma, Balantidium coli, Plasmodium, Toxoplasma*.

8. Sarcodina: pseudopods, Mastigophora: flagella, Ciliophora: cilia.

9. These organisms form spores as a natural part of their life cycle and have no real means of locomotion.

10. *Trichonympha* is an example of symbiosis, because both the *Trichonympha* and the termite benefit from the situation. In the case of the tapeworm, only the tapeworm benefits. The host is hurt by the situation.

11. Ciliates require so much energy that they must have a nucleus (called the macronucleus) devoted solely to metabolism. The other, smaller nucleus (the micronucleus) controls reproduction.

12. In conjugation between paramecia, there is a mutual exchange of DNA so that each paramecium gets new DNA. We learned in Module #2 that when bacteria conjugate, only one bacterium (the recipient) gets new DNA.

13. Spores are formed as a natural part of an organism's lifestyle. Cysts, however, are only formed in the case of life-threatening conditions. If those conditions do not exist, cysts will not be formed. Thus, the first group produced cysts. The second group produced spores, making them a part of phylum Sporozoa.

14. A euglena can either live on the dead remains of other organisms or it can produce its own food by photosynthesis. This combination of autotrophic and heterotrophic behavior is rather unique in God's Creation.

15. Phylum Chrysophyta contains the diatoms, which are responsible for most of the world's photosynthesis.

16. In the answers below, we list all of the phyla that apply. You only need to list one.

Food vacuole - purpose: store food, phyla: Sarcodina, Mastigophora, Ciliophora

Contractile vacuole - purpose: remove excess water, reducing pressure, phyla: Sarcodina, Mastigophora, Ciliophora

Flagellum - purpose: locomotion, phylum: Mastigophora, Pyrrophyta

Pellicle - purpose: retains cell shape, phyla: Mastigophora, Ciliophora

Chloroplast - purpose: stores chlorophyll, phylum: Chlorophyta or Mastigophora

Eyespot - purpose: detects light, phylum: Mastigophora

Cilia - purpose: locomotion, phylum: Ciliophora

Nucleus - purpose: contains DNA, phyla: all phyla in Protista

Oral groove - purpose: food intake and conjugation, phylum: Ciliophora

17. These deposits are called diatomaceous earth and are used as abrasives and filters.

18. A red tide is an algae bloom of dinoflagellates, which belong to phylum Pyrrophyta.

19. Phaeophyta and Rhodophyta

20. Members of phylum Phaeophyta have alginic acid (or just algin) in their cell walls. This is the thickening agent used in the foods listed.

SOLUTIONS TO THE STUDY GUIDE FOR MODULE #4

1. a. Extracellular digestion - Digestion that takes place outside of the cell

b. Mycelium - The part of the fungus responsible for extracellular digestion and absorption of the digested food

c. Hypha - Filament of fungal cells

d. Rhizoid hypha - A hypha that is imbedded in the material on which the fungus grows

e. Aerial hypha - A hypha that is not imbedded in the material upon which the fungus grows

f. Sporophore - Specialized aerial hypha that produces spores

g. Stolon - An aerial hypha that asexually reproduces to make more filaments

h. Haustorium - A hypha of a parasitic fungus which enters the host's cells, absorbing nutrition directly from the cytoplasm

i. Chitin - A chemical that provides both toughness and flexibility to a cell wall

j. Membrane - A thin covering of tissue

k. Fermentation - The anaerobic (without oxygen) breakdown of sugars into alcohol, carbon dioxide, and lactic acid.

l. Zygospore - A zygote surrounded by a hard, protective covering

m. Zygote - The result of sexual reproduction when each parent contributes half of the DNA necessary for the offspring

n. Antibiotic - A chemical secreted by a living organism that kills or reduces the reproduction rates of other organisms

2. Characteristics common to the majority of fungi were discussed in the section entitled "General Characteristics of Fungi." It was noted, however, that of the specialized hyphae, only rhizoid hyphae are common to all fungi.

Common to the majority of fungi	Present in only some
extracellular digestion	stolons (specialized hyphae)
chitin	caps and stalks (only mushrooms have them)
mycelia	sporangiophores (specialized hyphae with enclosed spores)
hyphae	haustoria (specialized hyphae)
cells (all living creatures have)	motile spores (only phylum Mastigomycota)
rhizoid hyphae	septate hyphae (many have non-septate hyphae)

3. Typically, we see only the fruiting body of a mushroom. Like an iceberg, that visible part is only a small fraction of the total mushroom, because the mycelium is the largest component of a mushroom.

4. Septate hyphae have cell walls to separate the cells while non-septate hyphae do not.

5. Rhizoid hyphae support the fungus and digest the food; a stolon asexually reproduces; a sporophore releases spores for reproduction; and a haustorium invades the cells of a living host to absorb food directly from the cytoplasm.

6. Stolons and sporophores are aerial. Aerial hyphae are not imbedded in the material upon which the fungus grows. In order to perform their jobs, rhizoid hyphae and haustoria must be imbedded in the material.

7. A sporangiophore produces its spores in an enclosure, a conidiophore does not.

8. Organisms in phylum Mastigomycota form motile spores, while organisms in phylum Amastigomycota form non-motile spores. Organisms in phylum Myxomycota resemble protozoa in their feeding stage and fungi in their reproductive stage. Imperfect Fungi contains all fungi that cannot be classified in Mastigomycota or Amastigomycota because we cannot identify a sexual stage of reproduction that produces spores.

9. A mushroom begins life as a small mycelium that grows from spores which have come from another mushroom. As the mycelium begins to grow, it might encounter the mycelium of another mushroom nearby. As the two mycelia begin to intertwine, their hyphae will sexually reproduce. Eventually, through some process that we do not understand, a group of the hyphae will form a complex web and enclose themselves in a membrane. When the hyphae are formed in the membrane, we say that the mushroom has reached the button stage of its existence. At that point, the hyphae begin filling with water quickly, and eventually the stipe and cap (the fruiting body) of the mushroom break through the membrane. The fruiting body of the mushroom releases its spores, which will grow into new mycelia if they land in suitable habitats.

10. The main difference is where they form their spores. Mushrooms form them on basidia that exist in the gills of the cap, puffballs produce them on basidia enclosed in a membrane, and shelf fungi produce them on basidia in pores on the fruiting body.

11. An alternate host is used by a parasitic fungus at some stage in its life. It is not the host that the fungus spends most of its life on, it is simply a temporary host that is necessary for a certain part of the fungus' development. Rusts use alternate hosts.

12. Yeast are best known for fermentation. They belong to class Ascomycetes.

13. In budding, the offspring stays attached to the parent until it has grown. In bacterial asexual reproduction, the offspring grows on its own.

14. There are many pathogenic fungi. You need only know two of them for the test:

1) rusts - crop damage
2) smuts - crop damage
3) ergot of rye (*Claviceps purpurea*) - death
4) *Cryphonectria parasitica* - chestnut blight
5) *Ophiostoma ulmi* - Dutch elm disease
6) *Phytophthora infestans* - late blight of potato

15. Members of class Basidiomycetes form spores on club-shaped cells known as basidia, while members of class Ascomycetes form their spores in sacs called asci. Members of class Zygomycetes produce zygospores.

16. Bread mold can asexually reproduce when a stolon elongates and eventually starts another mycelium. It can also asexually reproduce when an aerial hypha forms a sporophore (typically a sporangiophore). Sexually, bread mold reproduce when two mycelia form a zygospore.

17. If we do not know what its sexual mode of spore formation is, we place the fungus in phylum Imperfect Fungi.

18. If an antibiotic is used too much, resistant strains of the pathogen it is supposed to destroy can be formed.

19. Penicillin is extracted from a fungus in genus *Penicillium*.

20. In its feeding stage, a slime mold is a plasmodium. During that time, it resembles organisms from kingdom Protista.

21. Slime molds must have water to survive. Keep the habitat dry, and all slime molds will die.

22. Fungi participate in symbiosis by forming lichens and mycorrhizae. A lichen is a symbiotic relationship between a fungus and an algae. The algae produces food for both creatures via photosynthesis and the fungus supports and protects the algae. Mycorrhizae are symbiotic relationships between a fungus' mycelium and a plant's root system. The mycelium takes nutrients from the root while it collects minerals from the soil and gives it to the root.

23. A soredium is the specialized spore produced by most lichens. It contains spores for both the fungus and the algae.

SOLUTIONS TO THE STUDY GUIDE FOR MODULE #5

1. a. Matter - Anything that has mass and takes up space

b. Model - An explanation or representation of something that cannot be seen

c. Element - All atoms that contain the same number of protons

d. Molecules - Chemicals that result from atoms linking together

e. Physical change - A change that affects the appearance but not the chemical makeup of a substance

f. Chemical change - A change that alters the makeup of the elements or molecules of a substance

g. Phase - One of three forms - solid, liquid, or gas - which every substance is capable of attaining

h. Diffusion - The random motion of molecules from an area of high concentration to an area of low concentration

i. Concentration - A measurement of how much substance exists within a certain volume

j. Semipermeable membrane - A membrane that allows some molecules to pass through but does not allow other molecules to pass through

k. Osmosis - The tendency of a solvent to travel across a semipermeable membrane into areas of higher solute concentration

l. Catalyst - A substance that alters the speed of a chemical reaction but does not get used up in the process

m. Organic molecule - A molecule that contains only carbon and any of the following: hydrogen, oxygen, nitrogen, sulfur, and/or phosphorous

n. Biosynthesis - The process by which living organisms produce molecules

o. Isomers - Two different molecules that have the same chemical formula

p. Monosaccharides - Simple carbohydrates that contain three to ten carbon atoms

q. Disaccharides - Carbohydrates that are made up of two monosaccharides

r. Polysaccharides - Carbohydrates that are made up of several monosaccharides

s. Dehydration reaction - A chemical reaction in which molecules combine by ejecting water

t. Hydrolysis - Breaking down complex molecules by the chemical addition of water

u. Hydrophobic - Lacking any affinity to water

v. Saturated fat - A lipid made from fatty acids which have no double bonds between carbon atoms

w. Unsaturated fat - A lipid made from fatty acids that have at least one double bond between carbon atoms

x. Peptide bond - A bond that links amino acids together in a protein

y. Hydrogen bond - A strong attraction between hydrogen atoms and certain other atoms (usually oxygen or nitrogen) in specific molecules

2. In an atom, protons and neutrons cluster together at the center, which is called the nucleus. Electrons orbit around the nucleus.

3. The number of electrons (or protons) in an atom determines the vast majority of its characteristics.

4. When a number appears after an atom's name, it tells you the sum of protons and neutrons in the atom's nucleus.

5. An element contains all atoms that have the same number of protons (and therefore the same number of electrons), regardless of the number of neutrons. An atom is a single entity, determined by its number of protons, electrons *and* neutrons.

6. Since atoms have the same number of electrons and protons, there must be 32 electrons.

7. The subscripts after the elemental abbreviations tell you how many of each atom is in the molecule. Thus, there are 3 carbons, 8 hydrogens, and 1 oxygen, for a grand total of 12 atoms.

8. a. Molecule, because it has several atoms linked together
b. Atom, because it specifies number of neutrons and protons
c. Element, because it is by itself but does not specify number of neutrons and protons

9. Adding energy causes molecules to go from solid to liquid to gas. Thus, the liquid will turn into a gas. To turn it into a solid, you must take energy from it.

10. A semipermeable membrane should NOT be used. For diffusion to work, both solute and solvent must be able to travel across the membrane. Semipermeable membranes typically allow only solvent molecules to pass.

11. Since the water levels changed, that means solvent traveled from one side of the membrane to the other, but solute did not. This is osmosis, which requires a semipermeable membrane.

12. a. Reactants appear to the left of the arrow. The number to the left of the chemical formulas, however, do not describe the reactants. Instead, they tell you how many of each reactant molecule. Thus, the reactants are N_2 and H_2.
b. Products appear on the right side of the arrow. The product is NH_3.
c. There are 3 H_2 molecules in the reaction, because of the 3 to the left of H_2.

13. Photosynthesis is represented by:

$$6CO_2 + 6H_2O \rightarrow C_6H_{12}O_6 + 6O_2$$

In order for a plant to carry out photosynthesis, it needs CO_2, H_2O, energy from sunlight, and a catalyst like chlorophyll.

14. Reactions can also be sped up by increasing temperature.

15. Carbohydrates have carbon, hydrogen, and oxygen and no other elements. In addition, like water, they must have twice as many H's as O's. Only molecule d fits that bill.

16. Dehydration reactions build up these molecules, and hydrolysis reactions, providing the proper enzyme exists, can break them down.

17. An acid must contain an acid group ($\overset{\overset{\displaystyle O}{\|}}{C}-OH$). Only molecule c has one.

18. The pH scale measures the acidity or alkalinity of a solution. On this scale, 7 is neutral. Lower than 7 pH's are acidic, and higher than 7 are alkaline. The lower the pH the more acidic and the higher the pH the more alkaline.

19. Amino acids link together to make proteins, fatty acids link to glycerol to make lipids, and monosaccharides link together to make polysaccharides.

20. These proteins will not have the same properties. Not only the number and type but also the order of amino acids determine a protein's structure and function.

21. Enzymes are a special class of proteins that are used as catalysts.

22. The three basic parts of a nucleotide are the phosphate group, the sugar, and the base.

23. DNA stores information as a sequence of nucleotide bases, much like all of the English language can be stored as a sequence of dots and dashes in Morse code.

24. Hydrogen bonds between the nucleotide bases hold the two helixes of DNA together.

SOLUTIONS TO THE STUDY GUIDE FOR MODULE #6

1. a. Absorption - The transport of dissolved substances into cells

b. Digestion - The breakdown of absorbed substances

c. Respiration - The breakdown of food molecules with a release of energy

d. Excretion - The removal of soluble waste materials

e. Egestion - The removal of non-soluble waste materials

f. Secretion - The release of biosynthesized substances for use by other cells

g. Homeostasis - Maintaining the status quo in a cell

h. Reproduction - Producing more cells

i. Cytology - The study of cells

j. Cell wall - A rigid substance on the outside of certain cells, usually plant and bacteria cells

k. Middle lamella - The thin film between the cell walls of adjacent plant cells

l. Plasma membrane - The semipermeable membrane between the cell contents and either the cell wall or the cell's surroundings

m. Cytoplasm - A jelly-like fluid inside the cell in which the organelles are suspended

n. Ions - Substances in which at least one atom has an imbalance of protons and electrons

o. Cytoplasmic streaming - The motion of the cytoplasm which results in a coordinated movement of the cell's organelles

p. Mitochondria - The organelles in which nutrients are converted to energy

q. Lysosome - The organelle in animal cells responsible for hydrolysis reactions which break down proteins, polysaccharides, disaccharides, and some lipids

r. Ribosomes - Non-membrane-bound organelles responsible for protein synthesis

s. Endoplasmic reticulum - An organelle composed of an extensive network of folded membranes which perform several tasks within a cell

t. Rough ER - ER that is dotted with ribosomes

u. Smooth ER - ER that has no ribosomes

v. Golgi bodies - The organelles in which proteins and lipids are stored and then modified to suit the needs of the cell

w. Leucoplasts - Organelles that store starches or oils

x. Chromoplasts - Organelles that contain pigments used in photosynthesis

y. Central vacuole - A large vacuole that rests at the center of most plant cells and is filled with a solution which contains a high concentration of solutes

z. Waste vacuoles - Vacuoles that contain the waste products of digestion

aa. Phagocytosis - The process by which a cell engulfs foreign substances or other cells

bb. Phagocytic vacuole - A vacuole that holds the matter which a cell engulfs

cc. Pinocytic vesicle - Vesicle formed at the plasma membrane to allow the absorption of large molecules

dd. Secretion vesicle - Vesicle that holds secretion products so that they can be transported to the plasma membrane and released

ee. Microtubules - Spiral strands of protein molecules that form a rope-like structure

ff. Nuclear membrane - A highly-porous membrane that separates the nucleus from the cytoplasm

gg. Chromatin - Clusters of DNA and proteins in the nucleus

hh. Phospholipid - A lipid in which one of the fatty acid molecules has been replaced by a molecule which contains a phosphate group

ii. Passive transport - Movement of molecules through the plasma membrane according to the dictates of osmosis or diffusion

jj. Active transport - Movement of molecules through the plasma membrane (typically opposite the dictates of osmosis or diffusion) aided by a chemical process

kk. Isotonic solution - A solution in which the concentration of solutes is essentially equal to that of the cell which resides in the solution

ll. Hypertonic solution - A solution in which the concentration of solutes is greater than that of the cell which resides in the solution

mm. Plasmolysis - A collapse of the cell's cytoplasm due to lack of water

nn. Cytolysis - The rupturing of a cell due to excess internal pressure

oo. Hypotonic solution - A solution in which the concentration of solutes is less than that of the cell which resides in the solution

pp. Activation energy - Energy necessary to get a chemical reaction going

qq. Messenger RNA - The RNA that performs transcription

rr. Transcription - The process in which mRNA produces a negative of a strand of DNA

ss. Translation - The process by which proteins are formed in the ribosome according to the negative in mRNA

tt. Codon - A sequence of three nucleotides on mRNA that refers to a specific type of amino acid

uu. Anticodon - A three-nucleotide sequence on tRNA

2. The plasma membrane is composed of phospholipids, cholesterol, and proteins.

3. A phospholipid has two fatty acid molecules and a small molecule with a phosphate group, whereas a normal lipid just has 3 fatty acid molecules. This makes the phospholipid have a hydrophilic end, which the regular lipid does not.

4. Since the phospholipids have a hydrophilic end and a hydrophobic end, they always know how to reassemble.

5. Active transport requires energy from the cell, whereas passive transport does not. Thus, the active transport would slow down.

6. Since it died by implosion, the cell lost water. Water is lost by osmosis when the cell is in a solution which has a higher concentration of solutes than the inside of the cell. Thus, this was a hypertonic solution.

7. There are three stages: glycolysis, the Krebs cycle, and the electron transport system. Glycolysis results in a gain of 2 ATPs, the Krebs cycle 2, and the electron transport system 32. Thus, the electron transport system produces the most energy.

8. ATP supplies a package for the energy produced in cellular respiration. It releases its energy gently, so that the energy does not destroy the cell.

9. The only stage that does not require oxygen is glycolysis.

10. With no ADP, the cell will not be able to make ATP in which to store the energy from cellular respiration. Thus, the cell could make energy, but it could never use the energy!

11. Guanine and cytosine can bond together, as can adenine and thymine. In RNA, however, uracil replaces thymine. Thus when DNA has an adenine, RNA will have a uracil. When DNA has a thymine, RNA will have an adenine. When DNA has a cytosine, RNA will have a guanine, and when DNA has a guanine, RNA will have a cytosine. This makes the mRNA sequence:

a. cytosine, guanine, uracil, uracil, adenine, cytosine

b. It takes 3 nucleotides to code for an amino acid. Since this has 6, it will code for 2 amino acids.

12. a. transcription b. translation

13. This is tRNA, because only tRNA has anticodons.

14. They have different functions, but the main difference is that rough ER has ribosomes while smooth ER does not.

15. A leucoplast stores starches and oils; a chromoplast stores pigments. Since chlorophyll is a pigment used in photosynthesis, it is stored in a chromoplast.

16. The cell gains 2 ATPs in anaerobic respiration and 36 ATPs in aerobic respiration.

17. The lysosome performs hydrolysis which breaks down large molecules (like polysaccharides) into small molecules (like monosaccharides).

SOLUTIONS TO THE STUDY GUIDE FOR MODULE #7

1.

a. Genetics - The science that studies how characteristics get passed from parent to offspring

b. Genetic factors - The general guideline of traits determined by a person's DNA

c. Environmental factors - Those "non-biological" factors that are involved in a person's surroundings such as the nature of the person's parents, the person's friends, and the person's behavioral choices

d. Spiritual factors - The quality of a person's relationship with God

e. Gene - A section of DNA that codes for the production of a protein or a portion of protein, thereby causing a trait

f. Chromosome - A strand of DNA coiled around and supported by proteins, found in the nucleus of the cell

g. Mitosis - The duplication of a cell's chromosomes to allow daughter cells to receive the exact genetic makeup of the parent cell

h. Interphase - The time interval between cellular reproduction

i. Centromere - Constricted region of a chromosome and the point at which duplicate DNA strands attach themselves

j. Mother cell - A cell ready to begin reproduction, containing duplicate DNA and centriole

k. Karyotype - The figure produced when the chromosomes of a species during metaphase are arranged according to size

l. Diploid cell - A cell whose chromosomes come in homologous pairs

m. Haploid cells - Cells that have only one of each chromosome

n. Diploid chromosome number (2n) - The total number of chromosomes in a diploid cell

o. Haploid chromosome number (n) - The number of homologous pairs in a diploid cell

p. Meiosis - The process by which a diploid (2n) cell forms four gametes (n)

q. Gametes - Haploid cells (n) produced by diploid cells (2n) for the purpose of reproduction

r. Virus - A non-cellular infectious agent that has two characteristics:
(1) It has genetic material inside a protective protein coat
(2) It cannot reproduce itself

s. Antibodies - Specialized proteins that aid in destroying infectious agents

t. Vaccine - A weakened or inactive version of a virus that stimulates the body's production of antibodies which can destroy the virus

2. This would not mean that murders have no fault for what they do. Most genes only establish genetic trends. Environmental and spiritual factors affect the extent to which you follow those trends. Even if you have a genetic tendency to murder, the choices that you make can keep you from following that tendency.

3. Without the proteins in a chromosome, the chromosome would unravel.

4. Prophase, metaphase, anaphase, telophase

5. a. Notice how there are two distinct nuclei far apart from each other and the plasma membrane is beginning to constrict. This is telophase.

b. The chromosomes are still in the nucleus, but they are distinct. This means that they are ready to start mitosis. Thus, this is prophase.

c. The chromosomes are lined up on the equatorial plane. This is metaphase.

d. The chromosomes are pulling away from each other, but they are not far apart. Also, the plasma membrane has not started to constrict. This is anaphase.

6. Diploid number is the total number of chromosomes in the cell. Haploid number is the number of homologous pairs. If there are a total of 16 chromosomes, then there must be 8 pairs. The haploid number is 8.

7. Since haploid number is the number of pairs and diploid is the total number, then the diploid number is 18.

8. A gamete is haploid while a normal cell is diploid. This means that a gamete has only one chromosome from each homologous pair. A normal cell always has both members of each homologous pair.

9. prophase I, metaphase I, anaphase I, telophase I, prophase II, metaphase II, anaphase II, telophase II.

10. Meiosis II: It is essentially mitosis acting on two haploid cells.

11. In meiosis I, a single diploid cell splits into two haploid cells with duplicated chromosomes. Thus, there are 2 cells. Since they are haploid, they have one chromosome from each pair. Since there are 7 pairs, each cell has 7 chromosomes. The chromosomes are duplicated, because the purpose of meiosis II is to separate the duplicates from the originals.

12. In meiosis II, the two haploid cells have the duplicate chromosomes and their originals separated, producing a total of 4 haploid cells with no duplicated chromosomes. Thus, there are 4 cells, there are still 7 chromosomes in each, but the chromosomes are not duplicated.

13. Male gametes are called sperm, while female gametes are called eggs.

14. Animal males produce 4 useful gametes with each meiosis, while animal females produce only 1.

15. A polar body is a non-functional female gamete, because it is far too small to function properly. An egg is the one female gamete produced by meiosis that is large enough to function properly.

16. Sperm have flagella; thus, the male gamete can move on its own.

17. The lytic pathway is the way in which viruses reproduce, killing the cells of its host.

18. No virus is alive, because a virus cannot reproduce on its own. A virus also has no means of taking in nutrients and converting them into energy.

19. A vaccine is only good if you take it before getting infected, because it is meant to build up the antibodies that you need to fight the virus off before it overwhelms your body.

20. The virus in a vaccine cannot enter the lytic pathway. This makes it unable to reproduce and kill cells in the process.

SOLUTIONS TO THE STUDY GUIDE FOR MODULE #8

1.

a. True breeding - If an organism has a certain characteristic that is always passed on to all of its offspring, we say that this organism bred true with respect to that characteristic.

b. Allele - One of a pair of genes that occupies the same position on homologous chromosomes

c. Genotype - Two-letter set that represents the alleles an organism possesses for a certain trait

d. Phenotype - The observable expression of an organism's genes

e. Homozygous genotype - A genotype in which both alleles are identical

f. Heterozygous genotype - A genotype with two different alleles

g. Dominant allele - An allele that will determine phenotype if even one is present in the genotype

h. Recessive allele - An allele that will not determine the phenotype unless the genotype is homozygous with that allele

i.

1. The traits of an organism are determined by its genes.

2. Each organism has two alleles that make up the genotype of a given trait.

3. In sexual reproduction, each parent contributes ONLY ONE of its alleles to the offspring.

4. In each genotype, there is a dominant allele. If it exists in an organism, the phenotype is determined by that allele.

j. Pedigree - A diagram that follows a particular species' phenotype through several generations

k. Monohybrid cross - A cross between two individuals concentrating on only one definable trait

l. Dihybrid cross - A cross between two individuals concentrating on two definable traits

m. Autosomes - Chromosomes that do not determine the sex of an individual

n. Sex chromosomes - Chromosomes that determine the sex of an individual

o. Autosomal inheritance - Inheritance of a genetic trait not on a sex chromosome

p. Genetic disease carrier - A person who is heterozygous in a recessive genetic disorder

q. Sex-linked inheritance - inheritance of a genetic trait located on the sex chromosomes

r. Mutation - A radical chemical change in one or more alleles

s. Change in chromosome structure - A situation in which the chromosome loses or gains genes during meiosis

t. Change in chromosome number - A situation in which abnormal cellular events in meiosis lead to either none of a particular chromosome in the gamete or more than one chromosome in the gamete

2. a. This homozygous genotype is "YY," resulting in a phenotype of yellow peas.
   b. This heterozygous genotype is "Yy," resulting in a phenotype of yellow peas.
   c. This homozygous genotype is "yy," resulting in a phenotype of green peas.

3. Meiosis separates the two alleles.

4. One parent is homozygous dominant, so its genotype is "AA." The other is heterozygous, so its genotype is "Aa." The Punnett square looks like:

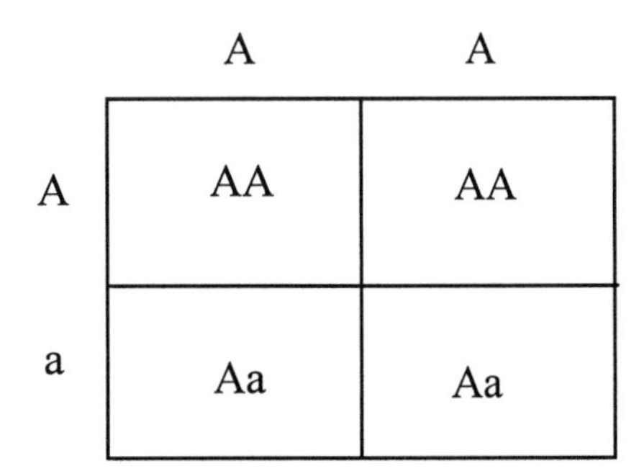

Thus, 50% of the offspring have the "AA" genotype and 50% have the "Aa" genotype. Since each offspring has at least one of the dominant allele, however, 100% have the axial flower phenotype.

5. Since the woman is heterozygous, her genotype is "Rr." The man cannot roll his tongue. Since the inability to roll your tongue is recessive, his genotype must be "rr." The resulting Punnett square is:

	R	r
r	Rr	rr
r	Rr	rr

Since having even one dominant allele allows you to be able to roll your tongue, 50% of the children will be able to roll their tongue.

6. Since the square that represents the male parent is filled, it means that he has a black coat. This means he has at least one dominant allele. One of the offspring is white-coated. The only way that can happen is for each parent to have at least one recessive allele. Thus, the genotype is "Bb."

7. Individuals 1 and 2 can tell us which allele is dominant. After all, the offspring have both phenotypes. This means that at least one of them is homozygous recessive. Thus, each parent must have the recessive allele. They both have no wings, but they must also carry the allele for wings, since one of their offspring has no wings. This means no wings ("N") is the dominant allele. Since they each must also have the recessive allele, 1 and 2 must have the "Nn" genotype. Individual 3 must have genotype "nn," because it has wings, which is the recessive trait. Since none of the offspring between 3 and 4 have recessive traits, 4 must have the "NN" genotype.

8. Since one parent is homozygous, its genotype is "SSYY." Since the other expresses both recessive alleles, it must be homozygous in the recessive alleles. Thus, its genotype is "ssyy." Both of these parents can only produce one type of gamete each. The one parent can only produce a *SY* allele and the other can only produce a *sy*. This gives us a 1x1 Punnett square.

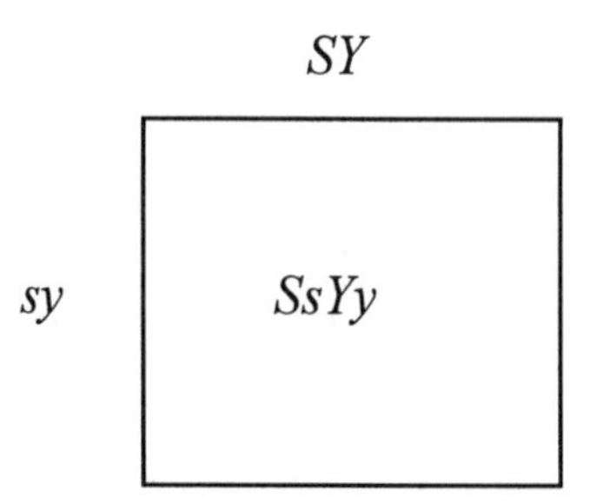

Since there is only one possible genotype, 100% of the offspring have the "SsYy" genotype and the smooth, yellow phenotype.

9. Since the parents are both heterozygous in each allele, their genotypes are "SsYy." There are 4 possible gametes: *SY, Sy, sY, sy*. The resulting Punnett square, then, is:

	SY	*Sy*	*sY*	sy
SY	*SSYY*	*SSYy*	*SsYY*	*SsYy*
Sy	*SSYy*	*SSyy*	*SsYy*	*Ssyy*
sY	*SsYY*	*SsYy*	*ssYY*	*ssYy*
sy	*SsYy*	*Ssyy*	*ssYy*	*ssyy*

smooth, yellow peas (genotypes SSYY, SsYy, SSYy, SsYY) 9 of 16 or 56.25 %
smooth, green peas (genotypes SSyy, Ssyy) 3 of 16 or 18.75 %
wrinkled, yellow peas (genotypes ssYY, ssYy) 3 of 16 or 18.75 %
wrinkled, green peas (genotype ssyy) 1 of 16 or 6.25 %

10. If the female is heterozygous, then her genotype is X^RX^r. Since the male is white-eyed, his genotype is X^rY. The resulting Punnett Square is:

	X^r	Y
X^R	X^RX^r	X^RY
X^r	X^rX^r	X^rY

Thus, 50% of the females will be white-eyed and 50% of the males will be white-eyed.

11. If the male were white-eyed, then the Punnett square would look like the one above, resulting in a white-eyed female. If the male were red-eyed, however, the resulting Punnett square:

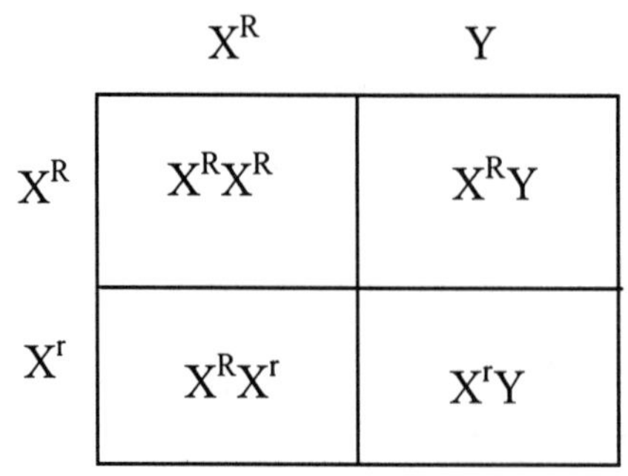

	X^R	Y
X^R	X^RX^R	X^RY
X^r	X^RX^r	X^rY

tells us that no white-eyed females are produced. Thus, the male's genotype is X^RY.

12. If a gamete has two alleles for the same trait, it must have two of the same chromosome. In the fertilization process, then, there will be three chromosomes. Thus, a genetic disorder from a change in chromosome number will result.

13. The genetic disorder must be recessive. Thus, the person can carry the trait but, as long as he or she has the dominant allele, the person will not have the disease.

14. Sex-linked disorders affect men more frequently than women. This is because men have only one allele in sex-linked traits.

15. Not all traits are determined completely by genetics. Most are also determined by environmental factors and (in the case of humans) spiritual factors. While the genetics are the same, the environmental and spiritual factors were probably different.

SOLUTIONS TO THE STUDY GUIDE FOR MODULE #9

1.

a. The immutability of species - The idea that each individual species on the planet was specially created by God and could never fundamentally change

b. Microevolution - The theory that natural selection can, over time, take an organism and transform it into a more specialized species of that organism.

c. Macroevolution - The hypothesis that the same processes which work in microevolution can, over eons of time, transform an organism into a completely different kind of organism

d. Strata - Distinct layers of rock

e. Fossils - Preserved remains of once-living organisms

f. Structural Homology - The study of similar structures (bones or organs, for example) in different species

2. He did most of his research while he was on board the *HMS Beagle*. True, it took him years after leaving the *Beagle* before publishing, but that was mostly because of his wife's urgings not to publish. Although he made most of his *observations* that led to his theory on the Galapagos archipelago, it was on the ship that he did most of the *work*.

3. No. Stories like that are simply lies.

4. Malthus believed in a constant struggle for survival. Without this idea of a constant struggle, Darwin would have never come up with the concept of natural selection.

5. Lyell came up with the idea that the present is the key to the past. He thought that the entire geological column could be explained by referring to the same processes that we see happening today. Darwin basically took that same idea and applied it to his hypothesis. He said that the variation we see in nature is due to the variation that occurs in reproduction, just like we see today.

6. Darwin dispelled the idea of the immutability of the species. By showing the evidence for microevolution, Darwin was able to show that species did change.

7. To go from a horse to a giraffe, there would need to be a lot added to the genetic code. Thus, this wild scenario is an example of macroevolution.

8. The fish remain fish; they have just varied their phenotype. Thus, this is variation within the genetic code, which is an example of microevolution.

9. In microevolution, the same genetic code exists throughout the change. The changes that occur are simply the result of variation within that genetic code. In order for macroevolution to occur, information must be added to the genetic code, essentially creating a new genetic code.

10.

Data Set	Summary
The geological column	This data is inconclusive as far as macroevolution is concerned. If you believe that the geological column was formed according to the speculations of Lyell, then it is evidence for macroevolution because it shows that life forms early in earth's history were simple and gradually got more complex. If you believe that the geological column was formed by natural catastrophe, then it is evidence against macroevolution. Since geologists have seen rock strata formed each way, it is impossible to tell which belief is scientifically correct.
The fossil record	This data is strong evidence against macroevolution. There are no clear intermediate links in the fossil record. The very few that macroevolutionists can produce are so similar to one of the two species they supposedly link, it is more scientifically sound to consider them a part of that species.
Structural homology	This data is strong evidence against macroevolution. The similar structures are not a result of inheritance from a common ancestor, because the similar structures are determined by quite different genes.
Molecular biology	This data is strong evidence against macroevolution. There are no evolutionary patterns in the sequences of amino acids of common proteins.

11. *Australopithecus afarensis* is supposed to be an intermediate link between man and ape. However, every bone that we have found of this creature indicates it is an ape. Thus, it is safest to assume that it is an ape. *Archaeopteryx* is supposed to link birds and reptiles, but once again the fossils tell us it is just a bird.

12. The most similar protein will be the one with the fewest difference in sequence. The protein in (a) has 5 amino acids different than the protein of interest, the protein in (b) has 4 differences and the one in (c) has 3. Thus, the protein in (c) is most similar.

13. A bacterium's cytochrome C should resemble a yeast's more than a kangaroo's does, because, according to evolutionists, the yeast evolved rather early after the bacterium, but the kangaroo came much, much later. In fact, however, the bacterium's cytochrome C sequence is *more similar to the kangaroo's* than it is to the yeast's!

14. Neo-Darwinism hoped to provide a mechanism by which information could be added to the genetic code of an organism. This was something Darwin's original hypothesis could not do.

15. Punctuated equilibrium attempts to explain away the fact that the fossil record is devoid of any intermediate links.

16. He would say that since the transition from species to species takes such a short amount of time, there is virtually no chance of an intermediate link being fossilized.

17. Structural homology and molecular biology still say that macroevolution (even by punctuated equilibrium) could not have happened.

SOLUTIONS TO THE STUDY GUIDE FOR MODULE #10

1.

a. Ecosystem - An association of living organisms and their physical environment

b. Ecology - The study of relationships among organisms in ecosystems

c. Primary consumer - An organism that eats producers

d. Secondary consumer - An organism that eats primary consumers

e. Tertiary consumer - An organism that eats secondary consumers

f. Ecological pyramid - A diagram that shows the biomass of organisms at each trophic level

g. Biomass - A measure of the mass of organisms within a region divided by the area of that region

h. Watershed - An ecosystem where all water runoff drains into a single river or stream

i. Transpiration - Evaporation of water from the leaves of a plant

j. Greenhouse effect - The process by which certain gases (principally water, carbon dioxide, and methane) trap heat that would otherwise escape the earth and radiate into space

2. If an insect not native to the U.S. were carried into the country through foreign fruits and vegetables, it could ruin the balance of the U.S. ecosystem.

3.

Organism	Possible Trophic Levels
Whale	primary consumer, secondary consumer
Sea turtle	primary consumer, secondary consumer
Phytoplankton	producer
Meran	primary consumer, secondary consumer
Ocean perch	secondary consumer
Zooplankton	primary consumer
Sea bass	secondary consumer, tertiary consumer
Shark	secondary consumer, tertiary consumer

4. a. The size of the rectangle indicates biomass. Also, the rectangles, in order, represent producers (bottom), then primary consumers, then secondary consumers, then tertiary consumers (top). Finally, we must look at percentage change, not absolute changes in the length. Thus, the primary and secondary consumers have the greatest disparity in biomass.

b. The smallest amount of energy is wasted where the biomass is as close to equal as possible. Thus, from producer to primary consumer wastes the least energy.

5. The clownfish and the sea anemone form a symbiotic relationship. The clownfish is protected by the sea anemone and it attracts food to the sea anemone. The goby and the blind shrimp have a symbiotic relationship in which the goby protects the blind shrimp, and the blind shrimp provides a home for the goby. Finally, the Oriental sweetlips and blue-streak wrasse form a symbiotic relationship in which the sweetlips gets it teeth cleaned by the wrasse and the wrasse gets food from the sweetlips' teeth.

6. Symbiosis seems to contradict the idea that organisms always battle for survival.

7. The ocean does not lose water because the land gets the excess water and it flows back into the ocean via surface runoff.

8. It transports nutrients.

9. If too many trees and plants are removed from a watershed, too many nutrients will flow into the river or stream, throwing off the ecosystem.

10. Oxygen is taken from the air principally by respiration and is restored principally by photosynthesis.

11. Oxygen is also removed from the air by fire, ozone formation, and the rusting of metals and minerals.

12. Oxygen is also restored by ozone destruction and water vapor destruction.

13. Carbon dioxide leaves the air by photosynthesis and by dissolving in the ocean.

14. Carbon dioxide enters the air via decomposition, fossil fuel burning, fire, and respiration.

15. Fossil fuel burning worries those who think that global warming is a problem, because it is a human-made way of adding more carbon dioxide to the air.

16. No, all measurable data indicate that any warming which did take place occurred before humans really started burning fossil fuels in earnest.

SOLUTIONS TO THE STUDY GUIDE FOR MODULE #11

a. Invertebrates - Animals that lack a backbone

b. Vertebrates - Animals that possess a backbone

c. Spherical symmetry - An organism possesses spherical symmetry if it can be cut into two identical halves by any cut that runs through the organism's center.

d. Radial symmetry - An organism possesses radial symmetry if it can be cut into two identical halves by any longitudinal cut through its center.

e. Bilateral symmetry - An organism possesses bilateral symmetry if it can only be cut into two identical halves by a single longitudinal cut along its center which divides it into right and left halves.

f. Epidermis - An outer layer of cells designed to provide protection

g. Mesenchyme - The jelly-like substance that separates the epidermis from the inner cells in a sponge

h. Collar cells - Flagellated cells that pump water into a sponge

i. Amebocytes - Cells in a sponge that perform digestion and transport functions

j. Gemmule - A cluster of cells encased in a hard, spicule-reinforced shell

k. Polyp - A sessile, tubular cnidarian with a mouth and tentacles at one end and a basal disk at the other

l. Medusa - A free-swimming cnidarian with a bell-shaped body and tentacles

m. Epithelium - Animal tissue consisting of one or more layers of cells that have only one free surface, because the other surface adheres to a membrane or other substance

n. Mesoglea - The jelly-like substance that separates the epithelial cells in a cnidarian

o. Nematocysts - Small capsules that contain a toxin which is injected into prey or predators

p. Testes - The organ that produces sperm

q. Ovaries - The organ that produces eggs

r. Anterior end - The end of an animal that contains its head

s. Posterior end - The end of an animal that contains the tail

t. Circulatory system - A system designed to transport food and other necessary substances throughout a creature's body

u. Nervous system - A system of sensitive cells that respond to stimuli such as sound, touch, and taste

v. Ganglia - Masses of nerve cell bodies

w. Hermaphroditic - Possessing both the male and the female reproductive organs

x. Regeneration - The ability to re-grow a missing part of the body

y. Mantle - A sheath of tissue that encloses the vital organs of a mollusk, secretes its shell, and performs respiration

z. Shell - A tough, multilayered structure secreted by the mantle. It is usually used for protection, but sometimes for body support

aa. Visceral hump - A hump that contains a mollusk's heart, digestive, and excretory organs

bb. Foot - A muscular organ that is used for locomotion and takes a variety of forms depending on the animal

cc. Radula - A organ covered with teeth that mollusks use to scrape food into their mouths

dd. Univalve - An organism with a single shell

ee. Bivalve - An organism with two shells

2. No. Eighteen of the 19 phyla in the animal kingdom are invertebrates (organisms with no backbones).

3. a. Bilateral, because it can only be cut into identical right and left halves

   b. Radial, because any up and down cut through the center makes two identical halves

   c. Spherical, because any cut through the middle makes two identical halves

4. Sponges get their prey by pumping water into themselves. The water brings algae, bacteria, and organic matter that sponges eat.

5. It contains spongin, because spongin is soft. Spicules make a sponge hard and prickly. These structures support the sponge.

6. When asexually reproducing, sponges use budding.

7. Amebocytes help digest and transport nutrients, they help carry waste to be excreted, they bring necessary gases such as oxygen to the cells, and they form the spicules or spongin.

8. A sponge produces gemmules during inclement times.

9. Hydra nematocysts are triggered with pressure, while the sea anemone's is triggered by an amino acid.

10. Cnidarians do not need these systems because their body walls are so thin that gases diffuse right through them.

11. Jellyfish spend part of their lives as polyps and the other part as medusas.

12. It must be in medusa form, because jellyfish can only reproduce sexually in medusa form.

13. Large coral colonies are called coral reefs.

14. *See Figure 11.7 for the answers.*

15. Earthworms bring minerals up from the lower parts of the soil and mix them with the nutrients at the top of the soil, which makes the soil fertile for plants. Their tunnels also allow oxygen to travel to the roots of a plant easier.

16. If the first earthworm feels slimy near the clitellum, this means that it is covered with a slime coat. Thus, the first one must have recently mated but not yet produced a cocoon.

17. The earthworm is hermaphroditic and the hydra can be as well. However, although a hydra can mate with itself, an earthworm cannot.

18. The earthworm will suffocate, because oxygen cannot travel through a dry cuticle.

19. Planarians do not need circulatory systems because the intestine is so highly-branched that all cells are near it, so they can get their food directly from the intestine.

20. Without complex nervous or digestive systems, it must not need to seek out and fully digest prey. The only way it can survive, therefore, is as a parasite.

21. When planarians asexually reproduce, they do so by regeneration.

22. a. Cnidaria b. Mollusca c. Porifera d. Platyhelminthes e. Annelida

SOLUTIONS TO THE STUDY GUIDE FOR MODULE #12

1. a. Exoskeleton - A body covering, typically made of chitin, that provides support and protection

b. Molt - To shed an old exoskeleton so that it can be replaced with a new one

c. Thorax - The body region between the head and the abdomen

d. Abdomen - The body region posterior to the thorax

e. Cephalothorax - A body region comprised of a head and a thorax together

f. Compound eye - An eye made of many lenses, each with a very limited scope

g. Simple eye - An eye with only one lens

h. Open circulatory system - A circulatory system that allows the blood to flow out of the blood vessels and into various body cavities so that the cells are in direct contact with the blood

i. Statocyst - The organ of balance in a crustacean

j. Gonad - A general term for the organ that produces gametes

2. Exoskeleton, body segmentation, jointed appendages, open circulatory system, and a ventral nervous system.

3. *See Figure 12.1 for solution*

4. *See Figure 12.3 for solution*

5. Blood collects in the pericardial sinus, and it enters the heart through one of three openings in the heart's surface. Each opening has a valve that closes when the heart is ready to pump. Once it absorbs the blood and closes these valves, the heart pumps blood through a series of blood vessels that are open at the end. These vessels dump directly into various body cavities. Gravity causes the blood to fall into the sternal sinus, where it is collected by blood vessels that are open at one end. Unlike the blood vessels that dump the blood into the body cavities, these vessels carry the blood back towards the pericardial sinus. On its way there, the blood is passed through the gills where it can release the carbon dioxide it has collected and pick up a fresh supply of oxygen. The blood also passes through green glands, which clean it of impurities and dump those impurities back into the surroundings. Once the blood has passed through the gills and the green glands, it then makes its way back to the pericardial sinus to begin the trip all over again.

6. It cleans the blood of impurities.

7. The maxillae are important. Without them, fresh, oxygen-rich water would not enter the gill chambers.

8. The injury gets sealed off to prevent bleeding, and then a new limb regenerates.

9. They are attached to the swimmerets.

10. They molt because their exoskeletons get too small for their growing bodies.

11. The antennules and antennae are responsible for taste and touch.

12. Four pairs of walking legs, two segments in body, no antennae, book lungs, four pairs of simple eyes.

13. No

14. No, some spiders that spin silk make trap doors with their silk or even fire the silk like a projectile.

15. The lung has many thin layers that look like the pages of a book.

16. Three pairs of walking (or jumping) legs, wings, three segments in body, one pair of antennae.

17. Insects do not need respiratory systems because of a complex network of tracheae that allow air to travel throughout the body.

18. The pupa stage only exists in complete metamorphosis.

19. membranous wings, scaled wings, leather-like wings, and horny wings.

20. a. Orthoptera

b. Hymenoptera

c. Diptera

d. Coleoptera

e. Lepidoptera

SOLUTIONS TO THE STUDY GUIDE FOR MODULE #13

1. a. Vertebrae - Segments of bone or some other hard substance that are arranged into a backbone

b. Notochord - A rod of tough, flexible material that runs the length of a creature's body, providing the majority of its support

c. Endoskeleton - A skeleton on the inside of a creature's body, typically composed of bone or cartilage

d. Bone marrow - A soft tissue inside the bone that produces blood cells

e. Axial skeleton - The portion of the skeleton that supports and protects the head, neck, and trunk

f. Appendicular skeleton - The portion of the skeleton that attaches to the axial skeleton and has the limbs attached to it

g. Closed circulatory system - A circulatory system in which the oxygen-carrying blood cells never leave the blood vessels

h. Arteries - Blood vessels that carry blood away from the heart

i. Capillaries - Tiny, thin-walled blood vessels that allow the exchange of gases and nutrients between the blood and cells

j. Veins - Blood vessels that carry blood back to the heart

k. Olfactory lobes - The regions of the brain that receive signals from the receptors in the nose

l. Cerebrum - The lobes of the brain that integrate sensory information and coordinate the creature's response to that information

m. Optic lobes - The regions of the brain that receive signals from the receptors in the eyes

n. Cerebellum - The lobe that controls involuntary actions and refines muscle movement

o. Medulla oblongata - The lobes that coordinate vital functions, such as those of the circulatory and respiratory system, and transport signals from the brain to the spinal cord

p. Internal fertilization - The process by which the male places sperm inside the female's body, where the eggs are fertilized

q. External fertilization - The process by which the female lays eggs and the male fertilizes them once they are outside of the female

r. Oviparous development - development that occurs in an egg which is hatched outside the female's body

s. Ovoviviparous development - development that occurs in an egg which is hatched inside the female's body

t. Viviparous development - development that occurs inside the female, allowing the offspring to gain nutrients and vital substances from the mother through a placenta

u. Anadromous - A lifecycle in which creatures are hatched in fresh water, migrate to salt water as adults, and then go back to fresh water in order to reproduce

v. Bile - A mixture of salts and phospholipids that aids in the breakdown of fat

w. Atrium - A heart chamber that receives blood

x. Ventricle - A heart chamber from which blood is pumped out

y. Ectothermic - Lacking an internal mechanism for regulating body heat

z. Hibernation - A state of extremely low metabolism

2. a. Class Amphibia b. Class Chondrichthyes c. Subphylum Cephalochordata d. Class Osteichthyes e. Subphylum Urochordata f. Class Agnatha

3. They all go through metamorphosis from larva to adult.

4. Bone is made up of cartilage that has been calcium-hardened. Thus, cartilage is more flexible and weaker than bone.

5. Capillaries have thin walls to allow for the diffusion of gases. Thus, this is, most likely, a capillary.

6. Red blood cells carry oxygen in the blood.

7. Hemoglobin gives red blood cells their color.

8. The cerebellum refines muscle movement. Creatures that have uncoordinated, jerky muscle movement have small cerebellums. Thus, amphibians have small cerebellums.

9. Vertebrates have enlarged lobes if the creature has a particular aptitude for the function controlled by the lobe. Since owls have good eyesight, their optic lobes are enlarged.

10. Fertilization is internal, because the female takes the sperm in to form the zygote, which is then encased in the egg. Development is oviparous, because the egg hatches externally.

11. The stronger the skeleton, the less flexible it is. Lampreys and rays both have cartilaginous skeletons, but the salmon is a bony fish. Thus, the salmon's skeleton is less flexible.

12. Atlantic salmon and many lamprey are anadromous.

13. The shark's most sensitive means of finding prey is its electrical field sensor.

14. The lateral line senses vibrations in the water. This alerts fish and sharks to movements in the water. Typically, sharks investigate the vibrations as possible food sources, while bony fish swim away from them in fear.

15. In both sharks and bony fish, the dorsal fins are used for balance in the water. In bony fish, the anterior dorsal fin is also a defensive weapon, because it is hard and sharp.

16. Rays tend to scurry around on the bottom of the ocean in the sand, while skates tend to swim a few feet above the bottom of the ocean.

17. *See Figure 13.10 for the answers.*

18.

Organ	Basic Function
Tongue	Taste
Pharynx	Sends food to esophagus
Gills	Exchange of carbon dioxide and oxygen between the water and the blood
Heart	Pumps blood
Liver	Makes bile for the digestion of fats
Gall bladder	Concentrates bile
Pyloric ceca	Secretes digestive enzymes and chemicals that break down food in stomach
Intestine	Digests food
Gonad	Reproduction
Anus	Expelling of undigested food
Brain	Controls nervous system
Esophagus	Sends food to stomach
Stomach	Stores and breaks down food
Spinal cord	Sends messages from brain to other parts of the body and vice-versa
Vertebral column	Support
Kidney	Cleans blood of waste products
Air bladder	Allows fish to change depths and float in water

19. *See Figure 13.11 for the answers.*

20.

Arteries	Veins	Neither
Efferent brachial arteries	Anterior cardial vein	Atrium
Dorsal aorta	Posterior cardial vein	Ventricle
Ventral aorta		Gills
Afferent brachial arteries		Kidney

21. Their endoskeleton is made mostly of bone.
Their skin is smooth with many capillaries and pigments. Amphibians do not have scales.
They usually have two pairs of limbs with webbed feet.
They have as many as four organs of respiration.
They have a three-chambered heart.
They are oviparous with external fertilization.

22. Frogs have smooth, wet skin and live near water. Toads have dry, warty skin and need not live near water.

23. The major respiratory organ for most amphibians is the skin.

SOLUTIONS TO THE STUDY GUIDE FOR MODULE #14

1. a. Botany - The study of plants

b. Perennial plants - Plants that grow year after year

c. Annual plants - Plants that live for only one year

d. Biennial plants - Plants that live for two years

e. Vegetative organs - The stems, roots, and leaves of a plant

f. Reproductive plant organs - The flowers, fruits, and seeds of a plant

g. Undifferentiated cells - Cells that have not specialized in any particular function

h. Xylem - A vascular tissue that carries substances upward in a plant

i. Phloem - Vascular tissue that carries substances downward in a plant

j. Leaf mosaic - The arrangement of leaves on the stem of a plant

k. Leaf margin - The characteristics of the leaf edge

l. Deciduous plant - A plant that loses its leaves before winter

m. Girdling - The process of cutting away a ring of inner and outer bark all the way around a tree trunk

n. Alternation of generations - A lifecycle in which sexual reproduction gives rise to asexual reproduction, which in turn gives rise to sexual reproduction

o. Dominant generation - In alternation of generations, the generation that occupies the largest portion of the lifecycle

p. Pollen - A fine dust that contains the sperm of seed-producing plants

q. Cotyledon - A "seed leaf" which develops as a part of the seed - it provides nutrients to the developing seedling and eventually becomes the first leaf of the plant.

2. Meristematic tissue will be anywhere that the plant is growing. After all, in order to grow, a plant's cells must carry on mitosis. The cells that perform mitosis are a part of the meristematic tissue.

3. The petiole attaches the leaf blade to the stem.

4. a. Whorled b. Alternate c. Opposite

5.

Letter	Shape	Margin	Venation
a.	Deltoid	Entire	Parallel
b.	Deltoid	Serrate	Pinnate
c.	Deltoid	Undulate	Pinnate
d.	Lobed	Dentate	Palmate
e.	Circular	Undulate	Pinnate (This is a tough one. You might think it's parallel, but there is actually a vein in the middle, from which the other veins sprout.)
f.	Chordate	Entire	Pinnate

6. The palisade mesophyll has the cells tightly packed, whereas the cells in the spongy mesophyll are loosely packed.

7. The guard cells control the opening and closing of the stomata.

8. The spongy mesophyll is typically on the underside of the leaf, and it is usually a lighter shade of green due to the fact that the photosynthesis cells are not as tightly packed there.

9. Carotenoids and anthocyanin.

10. No, a tree without an abscission layer cannot be deciduous. Remember, the abscission layer cuts off the flow of nutrients to the leaves, which causes them to stop doing photosynthesis, causing them to die. With no abscission layer, that will not happen and the tree will not lose its leaves in the winter.

11. The abscission layer is right between the stem and the petiole.

12. The four regions of a root are: the root cap, the meristematic region, the elongation region, and the maturation region. The undifferentiated cells are in the meristematic region.

13. a. This is from a dicot. The fibrovascular bundles do not have a face-like appearance; instead, they are characteristic of dicots.

b. This is from a monocot. The face-like characteristic of the fibrovascular bundles tells you this.

14. Woody stems have no limit to their growth because the cork cambium can always produce more bark. Thus, when the bark cracks, the inner parts of the stem are not exposed to the surroundings.

15. Mosses and ferns have the alternation of generations lifecycle. The dominant generation in the ferns is the sporophyte generation, while the mosses have the gametophyte generation as their dominant generation.

16. Since plants from phylum Bryophyta have no vascular tissue, there is no efficient way to transport nutrients throughout the plant. The plant must therefore stay small so that the nutrients need not travel far.

17. The plant must have a fibrous root system. If a root system does not go deeper than the height of a plant, it must spread out so that its total length is greater than that of the plant.

18. The female reproductive organ is the seed cone, and the male is the pollen cone.

19. The number of cotyledons produced in the seed is the fundamental difference between monocots and dicots.

20. In monocots, the venation is parallel while it is netted in dicots. The fibrovascular bundles are packaged differently in monocots and dicots. The root and stem structures are different. Typically, monocots have fibrous root systems whereas dicots have taproot systems. Finally, monocots usually produce flowers in groups of 3 or 6 while dicots produce flowers in groups of 4 or 5. The student need list only one of these.

SOLUTIONS TO THE STUDY GUIDE FOR MODULE #15

1. a. Physiology - The study of life processes that occur in the daily life of an organism

b. Nastic movement - Movement in a plant caused by changes in turgor pressure

c. Pore spaces - Spaces in the soil which determine how much water and air the soil contains

d. Loam - A mixture of gravel, sand, silt, and clay

e. Cohesion - The phenomenon that occurs when individual molecules are so strongly attracted to each other that they tend to stay together, even when exposed to tension

f. Translocation - The process by which organic substances move down the phloem of a plant

g. Hormones - Chemicals that affect the rate of cellular reproduction and the development of cells

h. Phototropism - A growth response to light

i. Gravotropism - A growth response to gravity

j. Thigmotropism - A growth rcsponse to touch

k. Perfect flowers - Flowers with both stamens and carpels

l. Imperfect flowers - Flowers with either stamens or carpels, but not both

m. Pollination - The transfer of pollen grains from the anther to the carpel in flowering plants

n. Seed - An ovule with a protective coating, encasing a mature plant embryo and a nutrient source

o. Fruit - A mature ovary which contains seeds

2. A plant uses water for photosynthesis, turgor pressure, hydrolysis, and transport. Since a plant can wilt without dying, turgor pressure can be ignored for a short time.

3. The first plant is using nastic movements and the second is using phototropism. Nastic movements refer to reversible, reasonably quick movements in a plant. Phototropism affects how a plant grows. The first plant is moving relatively quickly (for a plant) and is reversing that motion every day. That's nastic movement. The second plant grows towards the light. If the plant is not moved, this growth will not reverse itself. That's phototropism.

4. The cohesion-tension theory states that when water evaporates through the stomata in a plant's leaves, a deficit of water is created. This causes the water molecules just below those that evaporated to move up and take their place. Since water molecules like to stay together, however, the water molecules just below the ones that moved up also move up, in order to stay close. This causes a chain reaction, eventually causing water from the roots to move up into other parts of the plant.

5. Xylem cells need not be alive for xylem to do their job. Since we think that the cohesion-tension theory of water explains how water and dissolved nutrients travel up a plant, the xylem cells need not play an active role in the transport.

6. Phloem cells must be alive in order for the phloem to do their job, because the phloem cells take an active part in translocation.

7. Xylem contain water and dissolved minerals, while phloem contain organic substances.

8. Insectivorous plants do not really eat insects. They decompose the insects and use their raw materials for biosynthesis. Insectivorous plants produce their own food via photosynthesis just like other plants.

9. Vegetative reproduction leads to offspring with genetic codes which are identical to the parent. Sexual reproduction leads to offspring with genetic codes which are similar to, but not identical to, the parents' genetic codes.

10. The gardener must have grafted a limb from a tree which produces normal-sized apples to his crabapple tree.

11. *See Figure 15.4 for answers.*

12. The stamen is the male reproductive organ, and the carpel is the female reproductive organ.

13. There are at least 3 cells in a pollen grain. Two of them are sperm cells, and the other is a tube cell.

14. Typically, there are 7 cells in an embryo sac. Remember, the megaspore undergoes mitosis three times to make 8 nuclei. Then, the cell segments into six small cells and one large cell that has two nuclei. Two of these cells get fertilized. One becomes the zygote, and one becomes the endosperm.

15. The endosperm comes from the fertilization of the large, two-nucleus cell that is at the center of the embryo sac. It provides nutrition for the developing embryo.

16. Cotyledons either absorb the endosperm or aid in the transfer of nutrients from the endosperm to the embryo. This is how cotyledons provide a plant with nutrition before germination. After germination, they perform the first photosynthesis in the plant.

17. The fruit allows for the dispersal of seeds to places away from the parent.

18. There are many possible answers. The student needs at least 3:

wind, bees, beetles, birds, moths, or butterflies

SOLUTIONS TO THE STUDY GUIDE FOR MODULE #16

1. a. Amniotic egg - An egg in which the embryo is protected by a membrane called an amnion. In addition, the egg is covered in a hard or leathery covering.

b. Neurotoxin - A poison that attacks the nervous system, causing blindness, paralysis, or suffocation

c. Hemotoxin - A poison that attacks the red blood cells and blood vessels, destroying circulation

d. Endothermic - A creature is endothermic if it has an internal mechanism by which it can regulate its own body temperature, keeping it constant.

e. Down feathers - Feathers with smooth barbules but no hooked barbules

f. Contour feathers - Feathers with hooked and smooth barbules, allowing the barbules to interlock

g. Placenta - A structure that allows nutrients and gases to pass between the mother and the embryo

h. Gestation - The period of time during which an embryo develops before being born

i. Mammary glands - Specialized organs in mammals that produce milk to nourish the young

2.
- They are covered in tough, dry scales.
- They breath with lungs.
- They have a three-chambered heart with a ventricle which is partially divided.
- They produce amniotic eggs covered with a leathery shell. Most are oviparous, some are ovoviviparous.
- If they have legs, the legs are paired.

3. Reptiles are ectothermic, while birds and mammals are endothermic.

4. See Figure 16.1 for answers.

5. The yolk serves as nourishment for the developing embryo. The allantois allows the embryo to breathe, and the egg white destroys pathogens that can enter the egg through the porous shell.

6. They must both molt because their body covering is not living.

7. Reptile scales prevent water loss and insulate the reptile's body.

8. a. Squamata b. Rhynchocephalia c. Squamata d. Testudines e. Crocodilia f. Testudines

9.
- A bird is endothermic.
- A bird's heart has four chambers.
- A bird's mouth ends in a toothless bill.
- A bird is oviparous, laying an amniotic egg which is covered in a lime-containing shell.
- A bird's body is covered in feathers.
- A bird's skeleton is composed of porous, lightweight bones.

10. No. There are two orders of flightless birds.

11. If the blood sample has a mixture of oxygenated and de-oxygenated blood, it comes from an amphibian. If it has only one or the other, it comes from a bird. Remember, a bird's heart has four chambers, so oxygenated and de-oxygenated blood do not mix!

12. A bird's egg is harder, because it contains lime.

13. Only down feathers have no hooked barbules.

14. Contour feathers are used for flight, while down feathers are used for insulation.

15. When preening, a bird is actually oiling its feathers. The feathers need to be oiled regularly to keep the hooked barbules sliding freely along the smooth barbules and to keep the feathers essentially waterproof.

16. A bird's feathers molt in pairs, one at a time. This is different from arthropods and reptiles, who molt their outer covering all at once.

17. Flight engineers learned the proper structure of a wing from birds. They also learned how to make strong, hollow tubes from studying bird bones. Finally, they learned how to reduce wing turbulence from birds.

18. The amphibian's bone is heavier. Birds have air-filled cavities that make their bones lighter than other vertebrates' bones.

19.
- Mammals have hair covering their skin.
- Mammals reproduce with internal fertilization and are usually viviparous.
- Mammals nourish their young with milk secreted from specialized glands.
- Mammals have a four-chambered heart.
- Mammals are endothermic.

20. Underhair's main job is insulation.

21. We usually see the mammal's guardhair, because that's what's on top.

22. Any mammal from orders Monotremata or Marsupialia is non-placental. Thus, any one of the following: duck-billed platypuses, spiny anteaters, kangaroos, wallabies, koalas, opossums.

23. Offspring born after a long gestation period are more developed than those born after a short gestation period.

Tests

TEST FOR MODULE #2

1-4. Identify the structures that are left blank in the figure below:

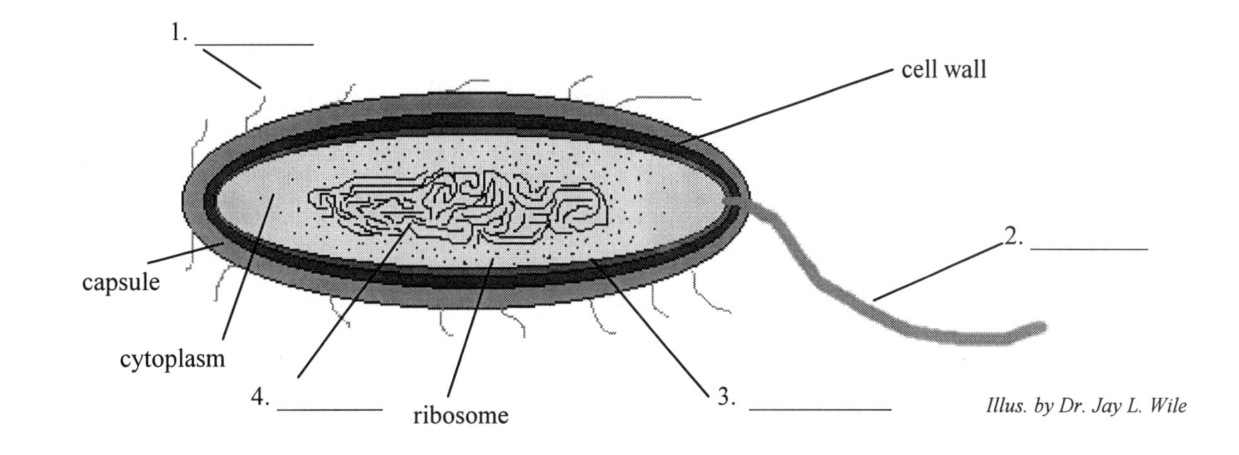

Illus. by Dr. Jay L. Wile

5. A bacterium suddenly cannot manufacture proteins. What component(s) of the cell is (are) not working?

6. If a bacterium is anaerobic, would you expect to find it floating at the top of a lake or deep in the muck at the bottom of the lake?

7. A bacterium is photosynthetic. Is it a decomposer?

8. A bacterium receives a new trait that it did not previously have. However, it did not participate in conjugation. How is this possible?

9. What kind of reproduction does the following schematic represent?

10. If a sample of food is dehydrated, what condition for bacterial growth are you removing from the food?

11. If a population of bacteria is in steady-state, does that mean no bacteria are dying?

12. What shape is a bacterium from the genus *Streptobacillus*.

13. A population of bacteria are very sensitive to light. They live in a lake that is in a cave, so they flourish. Slowly, however, a hole begins to erode in the cave's roof. As the days pass, light

begins to filter in and the cave starts to be dimly lit. The bacteria begin to die. While the erosion is taking place, however, two individual bacteria in the soil above ground fall into the lake. These bacteria cannot survive in the dimness of the poorly-lit cave, so they die immediately. Days later, the hole has opened up so that light floods the cavern. Nevertheless, the bacteria that were once dying are flourishing, with a population larger than they ever had before the erosion. What two things must have happened?

14. A bacterium is Gram-negative and needs light to survive. It lives in a habitat that is completely oxygen-free. To what phylum and class does it belong?

15. A Staphylococcus bacterium is Gram-positive and lives in an oxygen-free habitat. To what phylum and class does it belong?

TEST FOR MODULE #3

The classification groups within kingdom Protista

Subkingdoms	Phyla (in no particular order)
Algae	Mastigophora
Protozoa	Sarcodina
	Chlorophyta
	Chrysophyta
	Rhodophyta
	Sporozoa
	Pyrrophyta
	Ciliophora
	Phaeophyta

1. Define the following terms:

a. Thallus
b. Symbiosis
c. Vacuole

2. If an organism from kingdom Protista is heterotrophic, what subkingdom is it most likely in?

3. What phylum produces organisms whose remains are used as an abrasive in toothpaste? What is the generic name given to these organisms?

4. Some forms of algae have a chemical that can be used as a thickening agent in many consumer products. What is this chemical and in what phylum do these algae belong?

5. What phylum contains organisms that must have two nuclei? What are these nuclei called and what is the function of each?

6. Name one organism in kingdom Protista that is pathogenic. Name the malady that this organism causes.

7. If all diatoms were to suddenly go extinct, what would happen to the earth's atmosphere?

8. An organism forms a hard shell around itself in response to life-threatening conditions. If those life-threatening conditions had not occurred, it never would have behaved in such a way. Is this organism from phylum Sporozoa? Why or why not?

9. Two samples of cytoplasm from an amoeba are studied. The first is thin and watery, while the second is thick. Which sample was taken near the plasma membrane and which was taken from the center of the amoeba?

10. An organism from phylum Mastigophora cannot move. What organelle is not functioning?

11. An organism from phylum Ciliophora has no place to store its food once it takes the food in. What organelle is it missing?

12. If an organism in subkingdom Algae has chlorophyll, what organelle must it also have?

13. Even though kingdom Protista is mostly known for its single-celled creatures, there are two phyla that contain multicellular organisms. What phyla are they?

14. Classify each organism below into its proper subkingdom and phylum.

a. 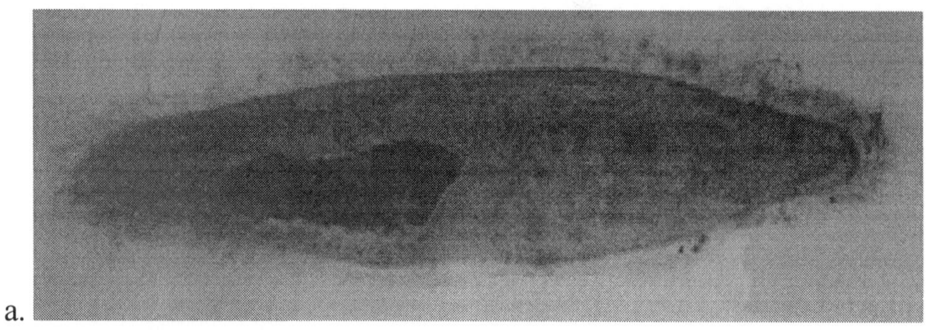

Photo by Kathleen J. Wile

b. 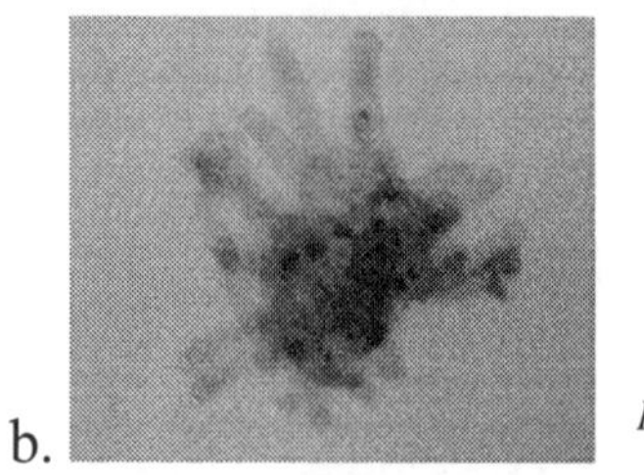

Photo by Kathleen J. Wile

c. 

Photo by Kathleen J. Wile

TEST FOR MODULE #4

1. Define the following terms:

a. Extracellular digestion
b. Rhizoid hypha
c. Stolon
d. Fermentation
e. Hypha

2. If a fungus forms haustoria, is it saprophytic or parasitic?

3. What kind of specialized hypha exists in all fungi?

4. What does chitin provide for a fungus?

5. Name two specialized hyphae that are not aerial.

6. Given the four phyla of kingdom Fungi: Mastigomycota, Amastigomycota, Imperfect Fungi, and Myxomycota, classify fungi with these characteristics:

a. produce motile spores
b. have no known sexual mode of reproduction
c. produce non-motile spores
d. resemble both protozoa and fungi

7. What part of the mushroom (the stipe, cap, or gill) holds the basidia?

8. Classify the following fungi into class Basidiomycetes, Ascomycetes, or Zygomycetes.

a. yeast
b. mushrooms
c. bread molds

9. Name two pathogenic fungi and the maladies that they cause.

10. A farmer notices that a certain crop grows much better in the presence of a certain fungi. What is the most likely explanation?

11. What useful medicine is produced by fungi in genus *Penicillium*? There is a general name for such medicines. What is that general name?

12. A biologist looks through a microscope at a single-celled life form. The microscope is not good enough to discern whether the cell is eukaryotic or prokaryotic. However, the biologist does see that the cell reproduces by budding. What is the most likely kingdom in which this organism belongs?

13. Why do slime molds appear in kingdom Protista in some biology books?

TEST FOR MODULE #5

1. Define the following terms:

a. Saturated fat
b. Physical change
c. Model
d. Isomers

2. A student reports on an atom that has 11 protons, 10 neutrons, and 10 electrons. What is wrong with the student's report?

3. Two atoms have the same number of neutrons but different numbers of protons. Do they belong to the same element?

4. The element sulfur is comprised of all atoms which have 16 protons. How many neutrons are in sulfur-34?

5. How many total atoms are in one molecule of $C_{22}H_{44}O$?

6. Identify the following as an atom, element, or molecule:

a. P   b. N_2 c. oxygen-16   d. PH_3

7. When you boil water, are you causing a chemical change or a physical change?

8. Two solutions of different solute concentration are separated by a membrane. After 1 hour, the solutions are of equal concentrations and the water levels are the same as when the experiment started. Is this an example of diffusion or osmosis?

9. In the following equation:

$$C_3H_8 + 5O_2 \rightarrow 3CO_2 + 4H_2O$$

Is CO_2 a product or a reactant? How many molecules of it are involved in the reaction?

10. What four things are necessary for photosynthesis?

11. If the following molecule is a carbohydrate, what is the value of x?

$$C_7H_xO_2$$

12. If you perform hydrolysis on a disaccharide, what kind of molecules will you get?

13. The pH of three solutions is measured. Solution A has a pH of 1.5; the pH of solution B is 7.1 and solution C has a pH of 13.2. Which solution or solutions is or are acidic?

14. Which of the following is an unsaturated fatty acid molecule?

a.
```
  H H H H H H H  O
  | | | | | | |  ||
H-C-C-C-C-C-C-C- C-OH
  | | | | | | |
  H H H H H H H
```

b.
```
  H  H  H  OH H H
  |  |  |  |  | |
H-C -C -C -C -C-C-OH
  |  |  |  |  || |
  OH OH OH H  O  H
```

c.
```
  H H H H     H  O
  | | | |     |  ||
H-C-C-C-C-C=C-C- C-OH
  | | | | | | |
  H H H H H H H
```

15. What determines the properties of a protein?

16. What part of the nucleotide is responsible for the way DNA stores its information?

17. What holds the two helixes in DNA together?

TEST FOR MODULE #6

1. Define the following terms:

    a. Cytoplasmic streaming

    b. Isotonic solution

    c. Cytolysis

    d. Phagocytosis

    e. Activation energy

    f. Plasmolysis

2. A cell produces a protein that will be used by other cells. When it ejects the protein, has it performed egestion, secretion, or excretion?

3. What lies in between the cell walls of a plant?

4. What is the difference between digestion and respiration?

5. Name at least two organelles that are involved in biosynthesis.

6. What organelle does rough ER have which smooth ER does not have?

7. Is a chloroplast an example of a leucoplast or a chromoplast?

8. Before a polysaccharide can be used in cellular respiration, to what organelle must it be sent?

9. If a cell's mitochondria stop working, can it perform any cellular respiration?

10. Which provides more energy: aerobic respiration or anaerobic respiration?

11. What stage in cellular respiration produces the most energy?

12. What gives the plasma membrane the ability to self-reassemble?

13. A strand of tRNA has the following nucleotide sequence:

adenine, uracil, guanine

What codon in mRNA attracts this strand of tRNA?

14. Part of a mRNA strand has the following nucleotide sequence:

    uracil, guanine, cytosine, cytosine, guanine, adenine, uracil, adenine, adenine

How many amino acids does this code for?

15. What DNA sequence produced the mRNA sequence given in problem #14?

TEST FOR MODULE #7

1. Define the following terms:

    a. Interphase
    b. Karyotype
    c. Diploid cell
    d. Gametes
    e. Virus

2. What factors besides genetics play a role in determining the characteristics of a person?

3. What is the purpose of the proteins in a chromosome?

4. Identify the stage of mitosis represented by each picture, then list the stages in the proper order.

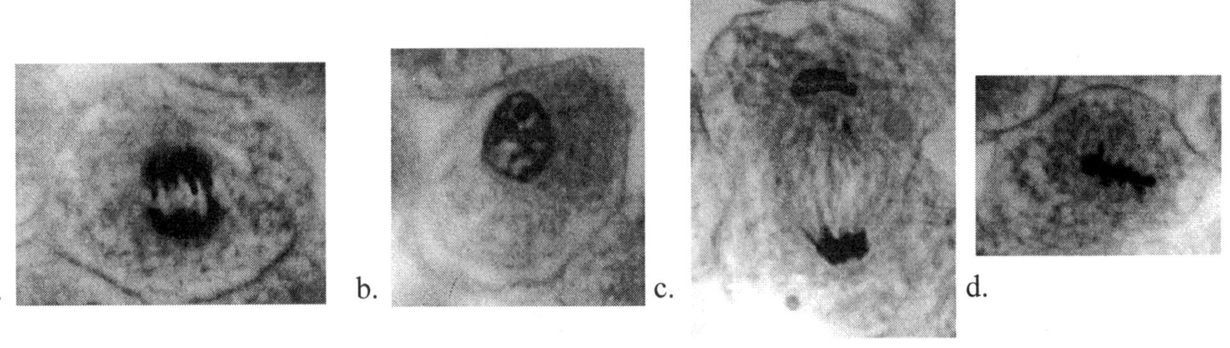

Photos by Kathleen J. Wile

5. A cell's DNA consists of 12 pairs of homologous chromosomes. What is its diploid number? What is its haploid number?

6. Three cells whose diploid number is 32 go through mitosis. How many cells result and how many total chromosomes are in each cell?

7. Three cells whose diploid number is 32 go through meiosis. How many cells result and how many total chromosomes are in each cell?

8. Which resembles mitosis most: meiosis I or meiosis II?

9. A haploid cell with duplicated chromosomes turns into two haploid cells with no duplicated chromosomes. Did the cell go through mitosis, meiosis I, or meiosis II?

10. A diploid cell with duplicated chromosomes turns into two diploid cells with no duplicated chromosomes. Did the cell go through mitosis, meiosis I, or meiosis II?

11. A single diploid cell goes through meiosis. Only one useful gamete is produced. Did this meiosis take place in a male or female?

12. There are two reasons a virus is not considered a living organism. Name one of those reasons.

13. What does a vaccine do to make a person immune to a virus?

14. What two different kinds of genetic material are found in viruses?

TEST FOR MODULE #8

1. Define the following terms:

a. True breeding
b. Allele
c. Homozygous genotype
d. Heterozygous genotype
e. Recessive allele
f. Monohybrid cross
g. Dihybrid cross
h. Autosomal inheritance

2. State the principles of Mendelian genetics.

3. In humans, the ability to roll one's tongue is a dominant genetic trait. If "R" represents this allele and "r" represents the recessive allele, what are the possible genotypes for a man who can roll his tongue?

4. For a given trait, how many alleles does a normal gamete have?

5. For a given trait, how many alleles does a non-gamete cell have?

6. a. The ability for a person to taste PTC is a dominant genetic trait ("T"), while the inability to taste PTC is recessive ("t"). If a man is heterozygous in that trait, what is his genotype?

b. If a woman cannot taste PTC, what is her genotype?

c. Draw the Punnett square for the children of this man and woman.

(Test continues on next page.)

7. The following pedigree is for humans, concentrating on the ability to roll their tongues. Since you already know that the ability to roll tongues is dominant and the inability is recessive, determine whether the filled circles and squares represent those who can or those who cannot roll their tongues.

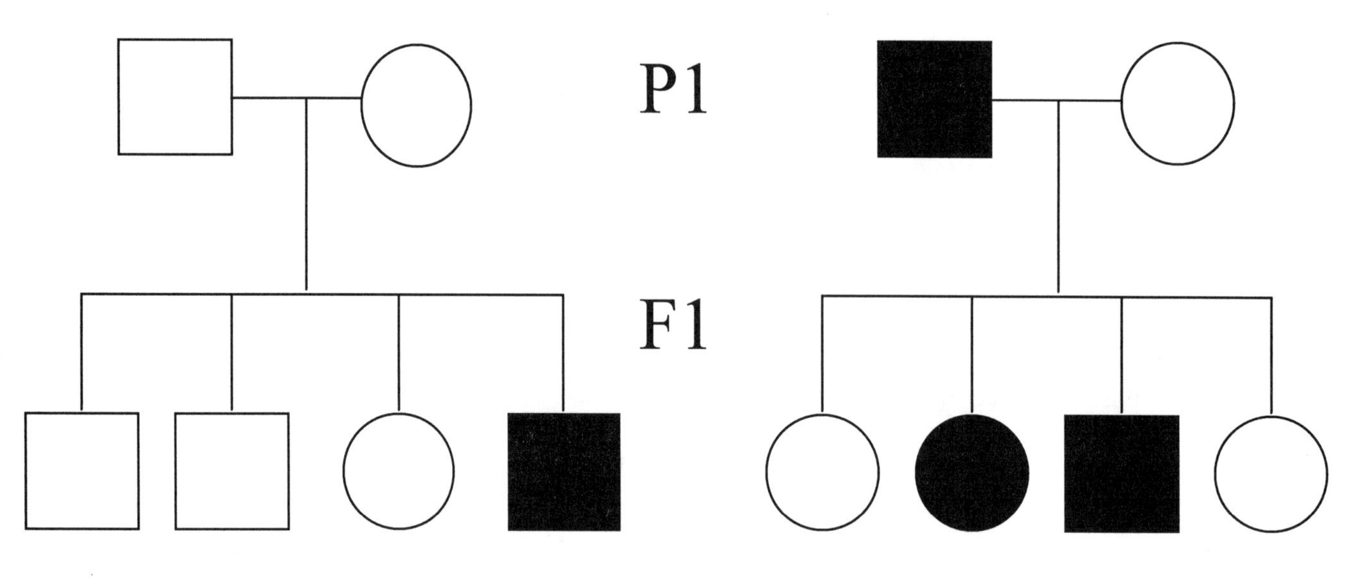

8. A pea plant is heterozygous in both the color of the pea produced ("Y" or "y") and the height of the plant ("T" or "t").

   a. What possible combinations of alleles exist in its gametes?

   b. Draw the resulting Punnett square for when this plant is self-pollinated.

   c. What are the possible phenotypes and their percent chances?

9. Why do recessive phenotypes in sex-linked traits show up in males significantly more often than in females?

10. Hemophilia is a sex-linked, recessive trait.

   a. Write the Punnett square for a non-hemophilic man having children with a woman who carries but does not have the disease.

   b. What percent of girls will have the disease?

   c. What percent of boys will have the disease?

11. Why would you not expect twins (who have identical DNA) to be identical in every way?

12. Name four means by which genetic disorders arise.

TEST FOR MODULE #9

1. Define the following terms:

   a. The immutability of species

   b. Microevolution

   c. Macroevolution

   d. Strata

   e. Fossils

   f. Structural Homology

2. What ship did Darwin travel on while working on his research?

3. Name the two ideas from other scientists that influenced Darwin in his work.

4. A horse breeder takes a horse that has won several races and breeds it with another horse that has won several races. These two horses produce 3 colts over their lifetime, and the breeder takes the fastest of these horses and breeds them again. After several generations of such a process, the breeder has produced a horse faster than any other in the world. Is this an example of microevolution or macroevolution?

5. If a colony of bacteria were to give rise to an amoeba, would this be an example of microevolution or macroevolution?

6. Of the four basic data sets discussed in this module, which ones provide conclusive evidence for macroevolution?

7. Of the four basic data sets discussed in this module, which ones provide conclusive evidence against macroevolution?

8. Some creationists say that all of the ideas set forth in Darwin's book, *The Origin of Species*, were wrong. Why is this not true?

9. The sequence of amino acids in the hemoglobin of a human is determined and compared to the sequence of amino acids in the hemoglobin of a rat and an ape. According to macroevolutionists, which of the two amino acids sequences should be closer to that of the human's?

10. How does the neo-Darwinist hypothesis differ from the hypothesis of Darwin?

11. What problem with both Darwin's original hypothesis and neo-Darwinism does punctuated equilibrium attempt to solve?

12. Consider the following amino acid sequences that make up a small portion of a protein:

a. Gly-Ile-Gly-Gly-Arg-His-Gly-Gly-Glu(NH_2)-Glu-Glu(NH_2)-Lys-Lys-Lys

b. Gly-Leu-Phe-Gly-Arg-Lys-Ser-Gly-Glu(NH_2)-Gly-Glu(NH_2)-Ala-Arg-Lys

c. Leu-Ile-Gly-Gly-Arg-His-Ser-Gly-Glu(NH_2)-Ala-Glu(NH_2)-Arg-Arg-Arg

Which protein would you expect to be the most similar to a protein with the following subset of amino acids?

Phe-Ile-Gly-Gly-Arg-His-Gly-Gly-Glu(NH_2)-Glu-Glu(NH_2)-Lys-Lys-Lys

TEST FOR MODULE #10

1. Define the following terms:

a.	Ecosystem	c.	Watershed	e. Greenhouse effect
b.	Biomass	d.	Transpiration	

Questions 2-4 refer to the following food web diagram:

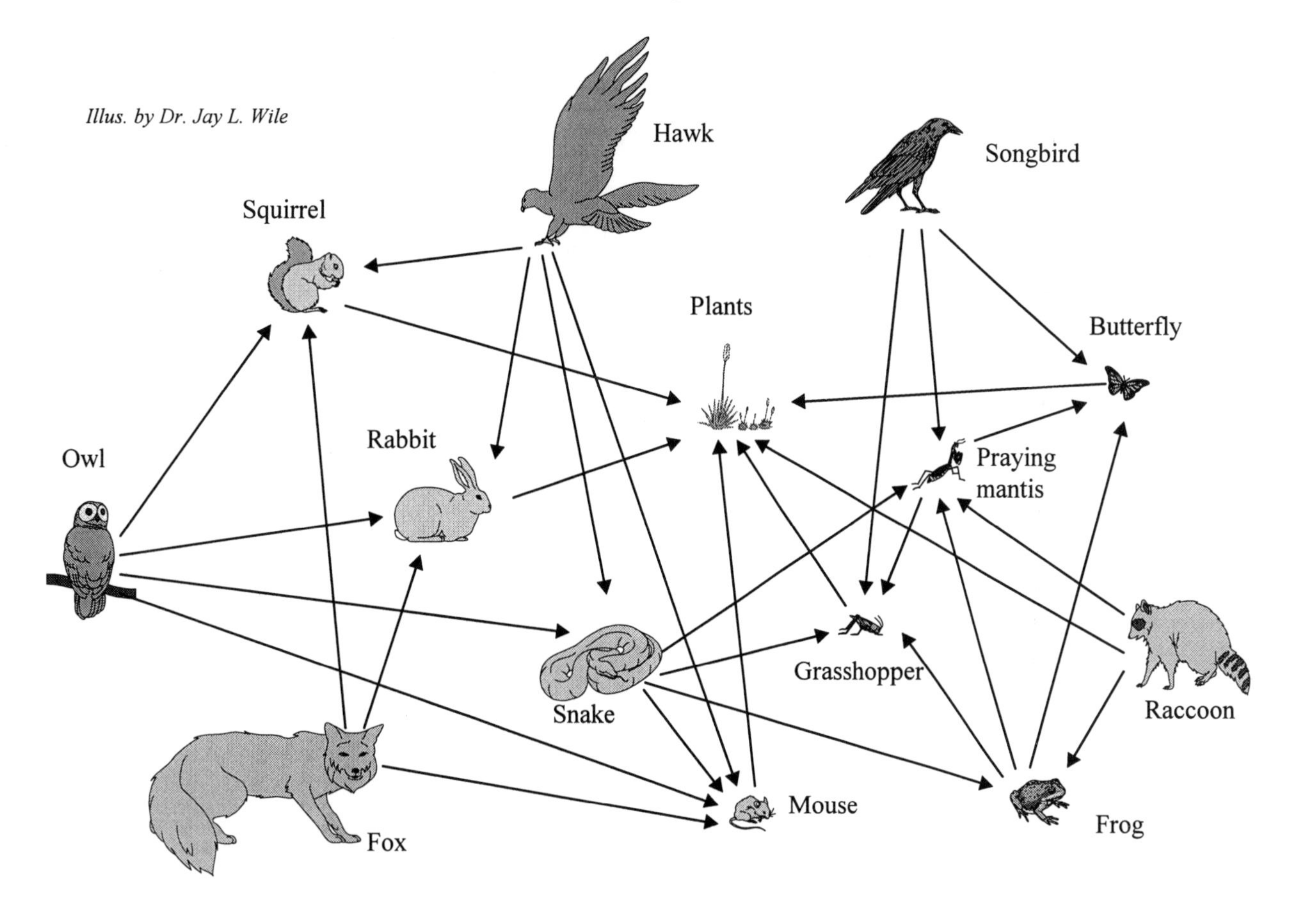

2. What is the most likely consequence of removing the snake and raccoon from this ecosystem?

3. What are the possible trophic levels of the owl?

4. What are the possible trophic levels of the frog?

5. Draw an ecological pyramid of an ecosystem that has twice as much biomass in producers as primary consumers, twice as much biomass of primary consumers as secondary consumers, and twice as much biomass of secondary consumers as tertiary consumers.

6. Name two of the symbiotic relationships that we learned about in this module and briefly discuss the role of each participant.

7. The stream flowing from a watershed has an algal bloom caused by too many nutrients in the water. What is the most likely explanation for this?

8. Does photosynthesis add oxygen to the air or remove it?

9. Besides photosynthesis, how can carbon dioxide be removed from the air?

10. If surface runoff did not occur in the water cycle of an ocean shore ecosystem, what would happen to the ocean?

11. Explain this statement: "The greenhouse effect is a good thing, but global warming is too much of a good thing."

12. Is global warming occurring today?

TEST FOR MODULE #11

1. Define the following terms:

a. Invertebrates

b. Vertebrates

c. Nematocysts

d. Posterior end

e. Hermaphroditic

f. Mantle

2. Place the following organisms into their proper phylum.

Phyla: Porifera, Cnidaria, Annelida, Mollusca, Platyhelminthes

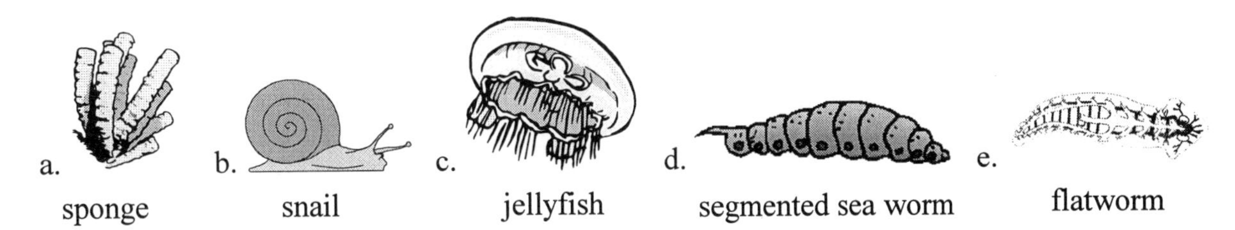

3. Are the following organisms univalve or bivalve mollusks?

4. What do sponges eat and where do they get their food?

5. What is the difference between a hard, prickly sponge and a soft sponge?

6. Which phylum of animals discussed in this module has nematocysts?

7. Identify the organs (a-g) in the following figure:

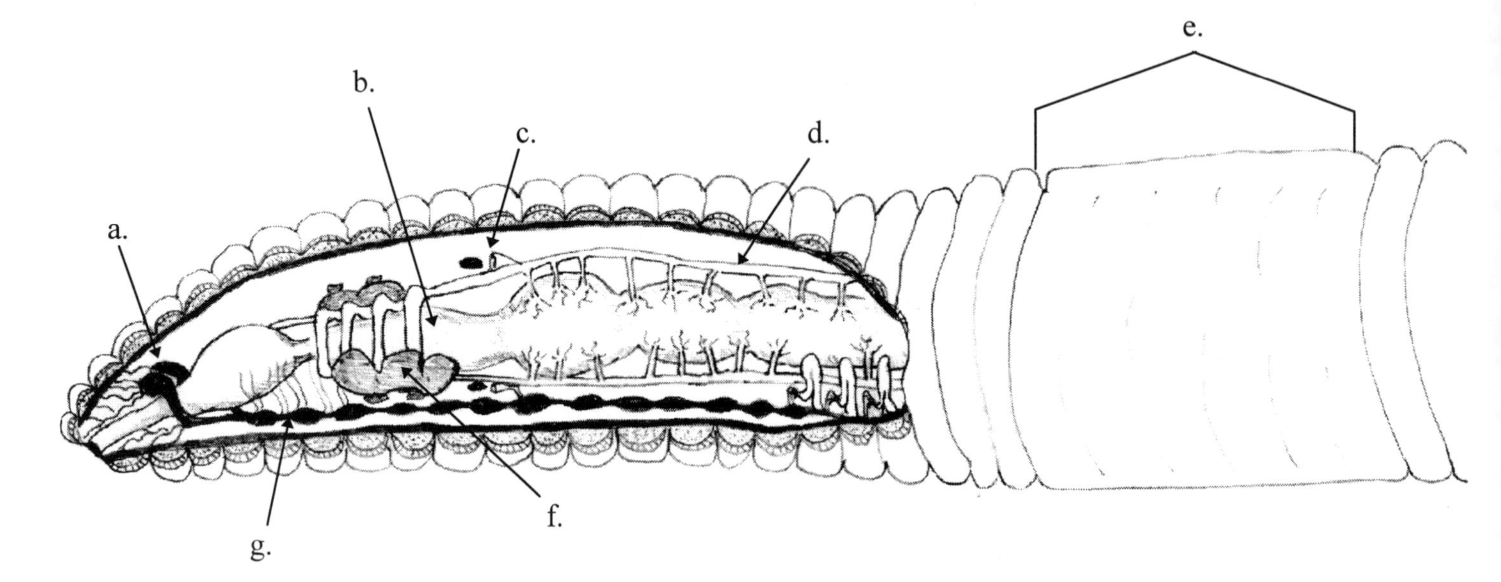

8. An earthworm's seminal vesicle is full. Has it mated yet?

9. What function does the cuticle perform in an earthworm?

10. Two members of phylum Platyhelminthes are studied. The first has a complex nervous system and the second does not. Which one is most likely the parasitic planarian?

11. What is the common mode of asexual reproduction among the organisms of phylum Cnidaria?

12. What is the common mode of asexual reproduction among the organisms of phylum Platyhelminthes?

TEST FOR MODULE #12

1. Define the following terms:

a. Exoskeleton
b. Simple eye
c. Open circulatory system
d. Statocyst
e. Gonad

2. Identify structures a-c in the following diagram:

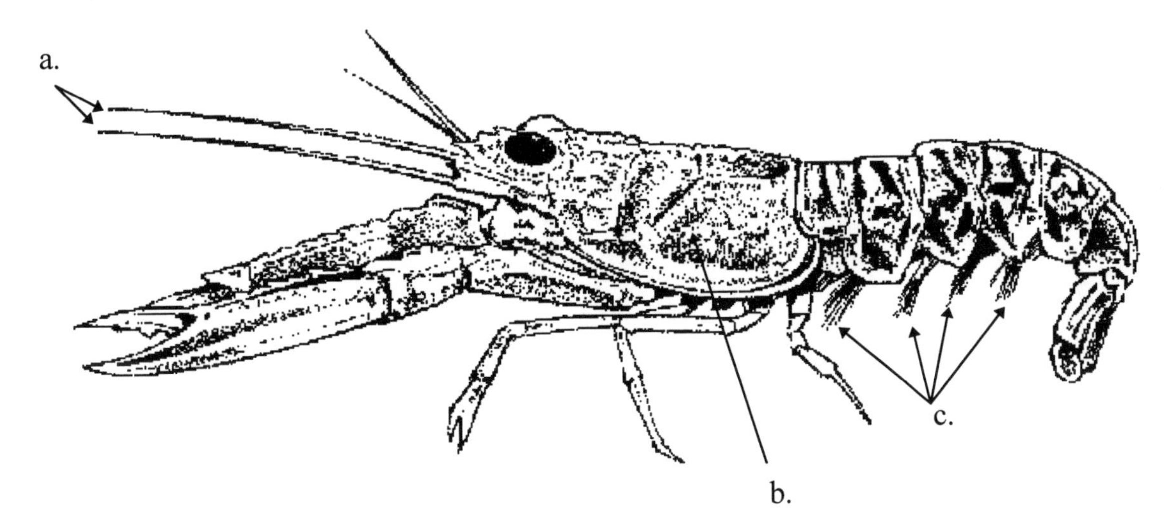

3. Identify organs a - g in the following diagram:

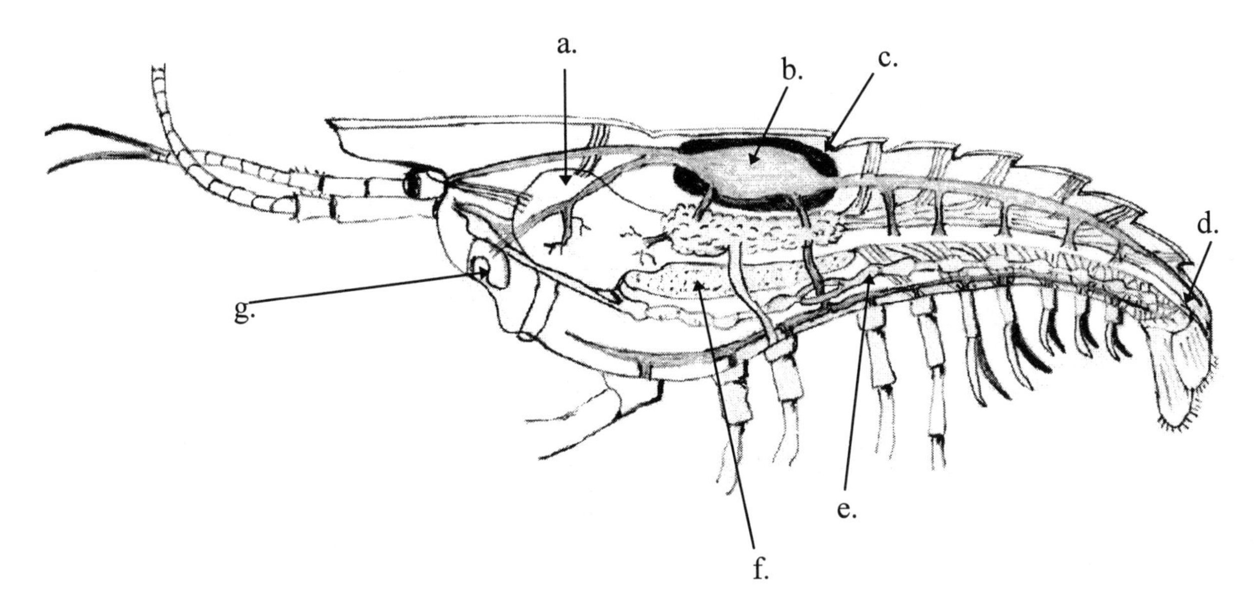

4. Why do arthropods molt?

5. What five characteristics set arachnids apart from the other arthropods?

6. What four characteristics set insects apart from the other arthropods?

7. What are the four stages of complete metamorphosis?

8. What takes the place of a respiratory system in insects?

9. If an insect has horny wings that protect membranous wings, to what order does it belong?

10. If an insect has leather-like wings that protect membranous wings, to what order does it belong?

TEST FOR MODULE #13

1. Define the following terms:

a. Bone marrow
b. Appendicular skeleton
c. Cerebrum
d. Cerebellum
e. Anadromous
f. Bile
g. Ectothermic
h. Hibernation

2. Name at least two chordates that have a larva stage as a part of their lifestyle.

3. Assign the following creatures to one of these classifications: subphylum Urochordata, subphylum Cephalochordata, class Agnatha, class Chondrichthyes, class Osteichthyes, class Amphibia

a. Ray b. Salmon c. Sea squirt d. Toad e. Salamander f. Lancelet g. Lamprey eel

4. If an animal's optic lobes are very small, what, most likely, is its weakest sense?

5. What protein allows red blood cells to perform their function?

6. An animal reproduces when the female lays eggs and the male then fertilizes them. The eggs are then left to develop and hatch. Is fertilization external or internal? What kind of development is this?

7. Which has the most flexible skeleton, a shark, a frog, or a lionfish?

Test Continues on Next Page

8. Identify the structures in this figure:

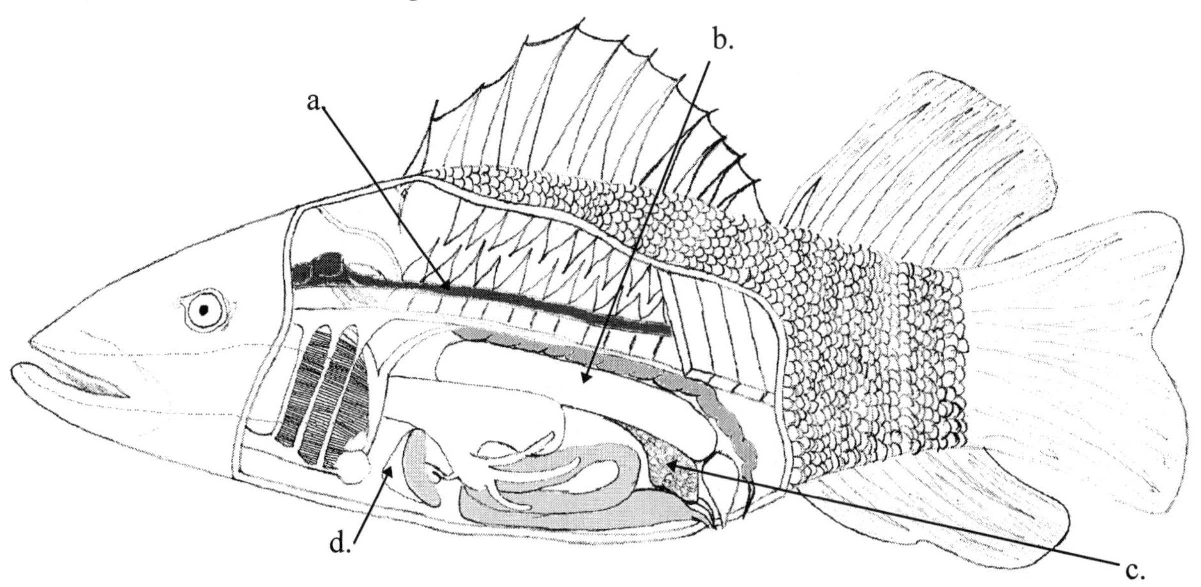

9. Identify the functions of the structures listed in problem #8

10. For each arrow in the figure below, indicate whether it is pointing to a vein or an artery.

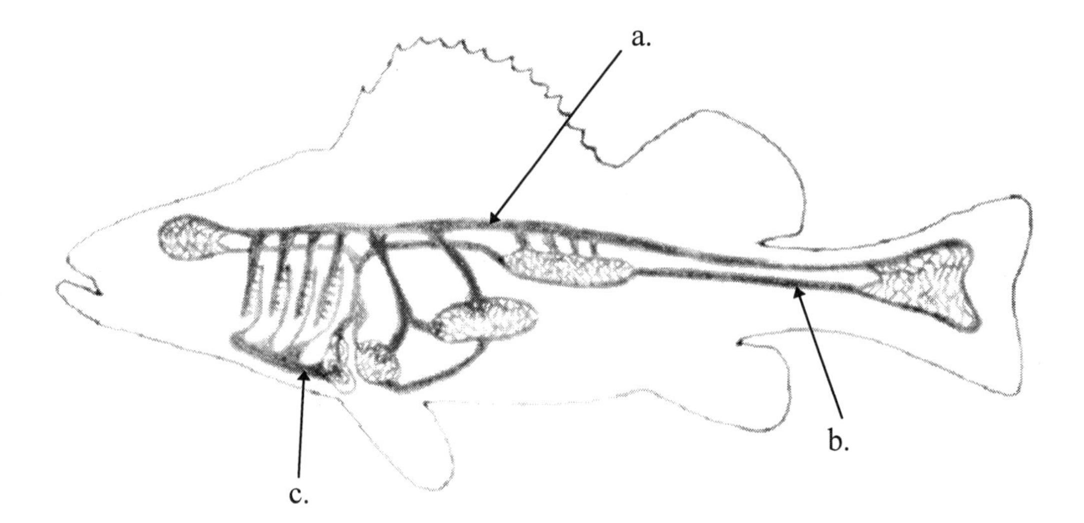

11. For each arrow in problem #10, indicate whether blood is flowing away from or to the heart.

12. For each arrow in problem #10, indicate whether the blood is oxygen-poor or oxygen-rich.

13. A member of Order Anura has smooth, wet skin. Is it a toad or frog?

14. What is the major respiratory organ for most amphibians?

TEST FOR MODULE #14

1. Define the following terms:

a. Vegetative organs

b. Reproductive plant organs

c. Undifferentiated cells

d. Xylem

e. Phloem

f. Leaf mosaic

g. Leaf margin

h. Deciduous plant

2. Determine the shape, margin, and venation of the following leaves:

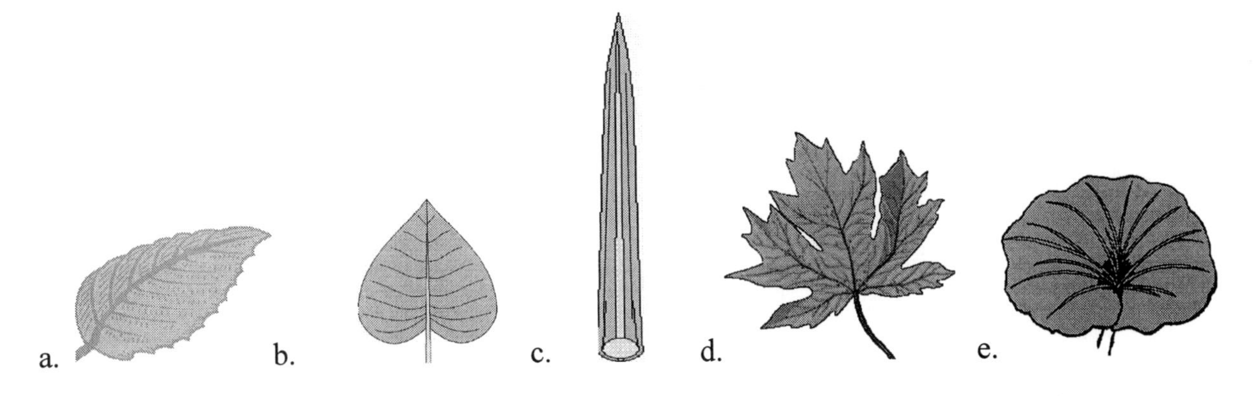

Illus. from the MasterClips collection

3. What function do the guard cells perform in a leaf?

4. A leaf has the spongy mesophyll on top. Which side of the leaf (top or bottom) will be the lighter shade of green?

5. What structure in a deciduous tree causes the leaves to die and fall off in the autumn?

6. What is anthocyanin, and what does it do to a leaf's color?

7. In which region of a root does the most growth take place?

8. Is this fibrovascular bundle from a monocot or a dicot?

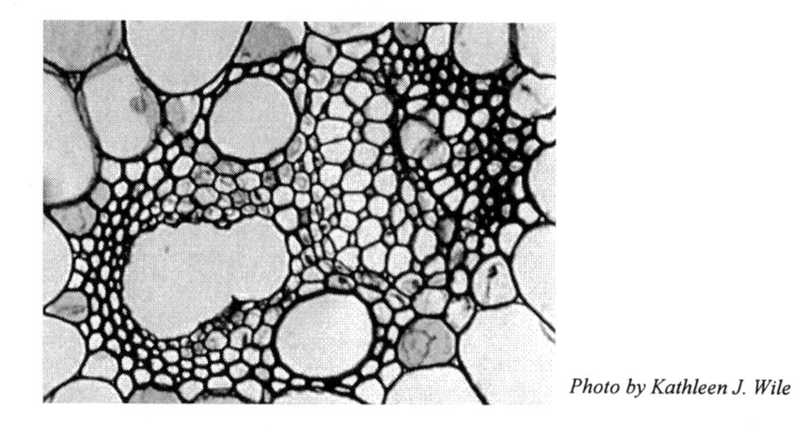

Photo by Kathleen J. Wile

9. Why is the bark of a tree often cracked?

10. A tree has seed cones and pollen cones. To which phylum does it belong?

11. A plant has a primary root that grows and grows without branching. What kind of root system is this?

12. Name two differences between monocots and dicots.

TEST FOR MODULE #15

1. Define the following terms:

    a. Physiology
    b. Nastic movement
    c. Pore spaces
    d. Loam
    e. Gravotropism
    f. Imperfect flowers

2. Name the four processes for which plants require water.

3. If a plant loses control of its stomata and they remain closed, will substances still flow through its xylem?

4. If a plant loses control of its stomata and they remain closed, will substances still flow through its phloem?

5. A biologist has a sample of fluid that came from a plant. If it is composed mostly of organic materials, did it come from the plant's xylem or phloem?

6. If an insectivorous plant cannot catch any insects, will it starve to death?

7. A gardener has two genetically identical plants. If one is the offspring of the other, did the parent propagate sexually or vegetatively?

8. Identify the structures in the figure below:

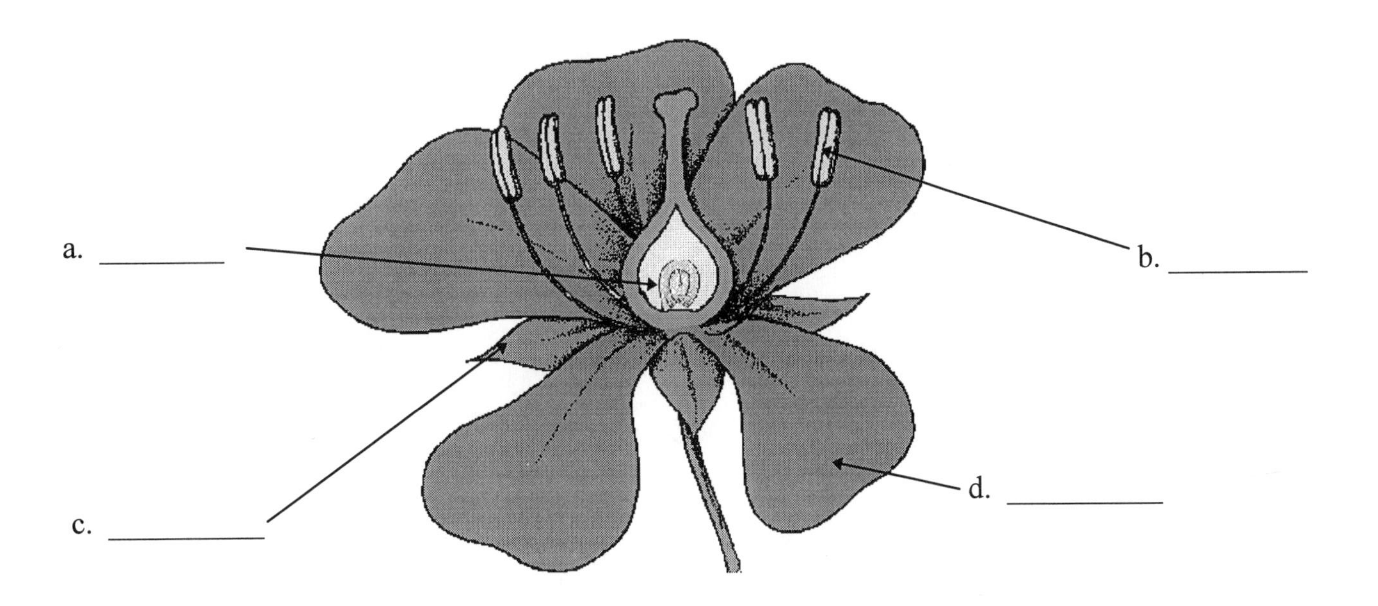

Match the following structure with the correct function.

9. Anther	a. Protects the sexual organs as they form
10. Ovule	b. Forms and releases pollen grains
11. Stigma	c. Holds the embryo sac
12. Sepal	d. Catches pollen grains

13. What function do cotyledons perform after germination?

14. What is the purpose of a fruit?

15. Name at least three ways in which pollen is transferred from the stamens of one flower to the carpels of another.

16. Which has more cells prior to fertilization, a pollen grain or an embryo sac?

17. When an embryo sac is fertilized, which cell is diploid: the zygote or the endosperm?

TEST FOR MODULE #16

1. Define the following terms:

    a. Amniotic egg
    b. Hemotoxin
    c. Endothermic
    d. Placenta
    e. Gestation

2. A vertebrate has a four-chambered heart, lays amniotic eggs with a lime-containing shell, and it has a light, porous skeleton. Is it a reptile, bird, or mammal?

3. A vertebrate has dry, tough scales, a three-chambered heart with a partial division in the ventricle, and it breathes with lungs. Is it a reptile, bird, or mammal?

4. A vertebrate has a four-chambered heart, hair, and nourishes its young with its own milk. Is it a reptile, bird, or mammal?

5. These are the reptile orders that contain currently living reptiles:

Crocodilia, Testudines, Rhynchocephalia, Squamata

Place the following types of reptiles into their appropriate order:

a. tuataras   b. lizards   c. crocodiles   d. turtles

6. A vertebrate is endothermic. Is it possible that the vertebrate is a reptile?

7. Which kind of feather has hooked barbules: contour feathers or down feathers?

8. You find an egg in your back yard. It is covered in a soft, leathery shell. Is it a reptile egg or a bird egg?

9. A bird's feathers become inflexible because the hooked barbules do not slide easily on the smooth barbules. What should the bird do to fix this problem?

10. A mammal lives in a cold climate and needs a lot of insulation. Will its guardhair or underhair be thicker than the average mammal's?

11. Two species of cat are very similar in appearance. The first species gives birth to offspring that have no hair and whose eyes are closed. The second species gives birth to offspring that have a full coat of hair and whose eyes are open. Which species has the shorter gestation period?

12. Identify the structures in the amniotic egg pictured below:

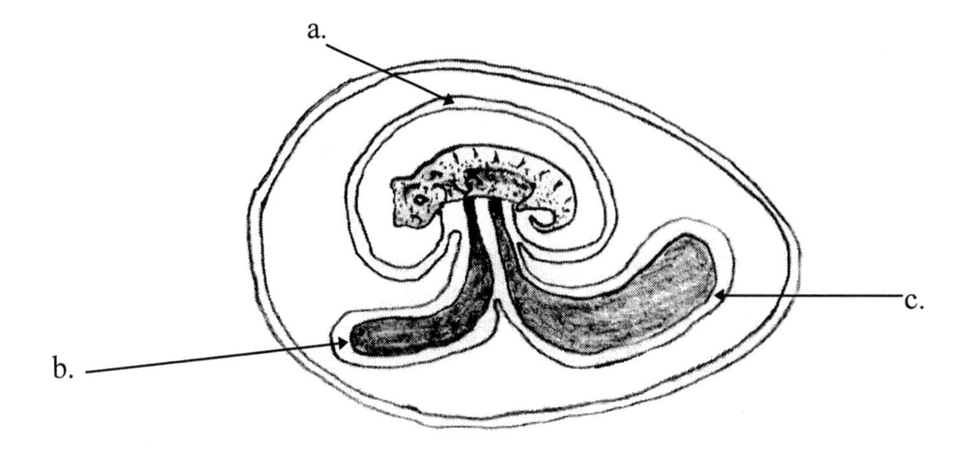

Solutions To The

Tests

SOLUTIONS TO THE TEST FOR MODULE #1

1. a. Heterotrophs - Organisms that depend on other organisms for food

   b. Mutation - An abrupt and marked difference between offspring and parent

   c. Theory - A hypothesis that has been tested with a significant amount of data

   d. Photosynthesis - The process by which a plant uses the energy of sunlight and certain chemicals to produce its own food. Oxygen is often a by-product of photosynthesis.

   e. Prokaryotic cell - A cell that has no distinct, membrane-bound organelles

2. heterotroph

3. It will not be able to extract energy from the surroundings and convert it into energy that sustains life.

4. A. Reproduce sexually
   B. Reproduce asexually
   C. Experiencing mutations

5. The two organisms from the same genus will probably have the most similarity.

6. Different phyla

7. It survived because flawed experiments seemed to confirmed it.

8. Fungi

9. Plantae

10. Monera

11. a. Kingdom: Plantae
    Phylum: Anthophyta
    Class: Dicotyledoneae

    b. Kingdom: Animalia
    Phylum: Chordata
    Class: Aves

SOLUTIONS TO THE TEST FOR MODULE #2

1. Pilus

2. Flagellum

3. Plasma Membrane

4. DNA

5. The ribosomes

6. Deep in the muck, because there is little or no oxygen down there.

7. No, photosynthetic bacteria are autotrophs.

8. It participated in transformation, the other means by which bacteria get new traits.

9. Asexual

10. Moisture

11. No, bacteria are simply reproducing as quickly as others are dying.

12. Since the name ends in bacillus, it is rod-shaped.

13. A bacterium must have participated in transformation so that the ability to live in light was absorbed from one or both of the dead bacteria that fell into the lake. Then, because there are so many bacteria present, that bacterium must have participated in conjugation with several others. NOTE: If transformation is present but conjugation is not, give half credit.

14. Gram negative bacteria are in phylum Gracilicutes. Anaerobic bacteria in this phylum are in class Anoxyphotobacteria.

15. Gram positive bacteria are in phylum Firmicutes. Class Firmibacteria contains those Gram positive bacteria that are coccus-shaped. The oxygen-free information is irrelevant, because respiration only matters in classifying Gram negative bacteria.

SOLUTIONS TO THE TEST FOR MODULE #3

1. a. Thallus - The body of a plant-like organism that is not divided into leaves, roots, or stems

b. Symbiosis - Two or more organisms of different species living together so that each benefits from the other

c. Vacuole - membrane bound "sac" within a cell

2. Protozoa

3. These organisms come from phylum Chrysophyta and are commonly called diatoms.

4. The chemical is alginic acid (algin) and it comes from members of phylum Phaeophyta.

5. Members of phylum Ciliophora have two nuclei, the macronucleus and the micronucleus. The macronucleus controls the organism's metabolism while the micronucleus controls reproduction.

6. There are several. Any of the following are correct answers:

Entamoeba histolytica causes dysentery
Trypanosoma causes African sleeping sickness
Balantidium coli causes dysentery
Plasmodium causes malaria
Toxoplasma causes toxoplasmosis

7. The oxygen supply would dwindle away.

8. No. Sporozoa form spores as a natural part of their lifestyle.

9. The first is taken near the plasma membrane. The second is taken near the center of the amoeba.

10. the flagellum

11. a food vacuole

12. a chloroplast

13. Phaeophyta and Rhodophyta

14. a. Subkingdom Protozoa, phylum Ciliophora
b. Subkingdom Protozoa, phylum Sarcodina
c. Subkingdom Algae, phylum Chlorophyta

SOLUTIONS TO THE TEST FOR MODULE #4

1. a. Extracellular digestion - Digestion that takes place outside of the cell

   b. Rhizoid hypha - A hypha that is imbedded in the material on which the fungus grows

   c. Stolon - An aerial hypha that asexually reproduces to make more filaments

   d. Fermentation - The anaerobic (without oxygen) breakdown of sugars into alcohol, carbon dioxide, and lactic acid.

   e. Hypha - Filament of fungal cells

2. parasitic

3. rhizoid hyphae

4. toughness and flexibility

5. rhizoid hyphae, haustoria

6. a. Mastigomycota b. Imperfect Fungi c. Amastigomycota d. Myxomycota

7. The gills (You can count cap as partial credit if you want, but the gills hold the basidia. The cap holds the gills.)

8. a. Ascomycetes b. Basidiomycetes c. Zygomycetes

9. There are many pathogenic fungi. Any two of these will suffice:

1) rusts - crop damage
2) smuts - crop damage
3) ergot of rye (*Claviceps purpurea*) - death
4) *Cryphonectria parasitica* - chestnut blight
5) *Ophiostoma ulmi* - Dutch elm disease
6) *Phytophthora infestans* - late blight of potato

10. Most likely, the crop and the fungus form a mycorrhizae together.

11. Penicillin: the general name for such a medicine is "antibiotic."

12. Fungi. It is most likely a yeast.

13. Because their classification is in some dispute. They resemble members of kingdom Protista during their feeding stage and members of Fungi during their reproductive stage. Thus, they could belong to either kingdom.

SOLUTIONS TO THE TEST FOR MODULE #5

1. a. Saturated fat - A lipid made from fatty acids which have no double bonds between carbon atoms

b. Physical change - A change that affects the appearance but not the chemical makeup of a substance

c. Model - An explanation or representation of something that cannot be seen

d. Isomers - Two different molecules that have the same chemical formula

2. An atom has to have the same number of protons as electrons.

3. No, to belong to the same element, the number of protons must be the same

4. 18

5. 67

6. a. element b. molecule c. atom d. molecule

7. physical change - all phase changes are physical changes

8. diffusion

9. product, 3

10. carbon dioxide, water, energy from sunlight, and a catalyst

11. Remember, a carbohydrate has the same ration of H's to O's as water (2:1). Thus, if there are 2 O's, there must be 4 H's.

12. monosaccharides

13. The pH scale says that pH's below 7 are acidic. The lower the pH, the more acidic. Thus, A is the only acidic one.

14. Acids must have an acid group. Thus, only A and C are acids. Unsaturated means there is at least one double bond between carbons. This tells us that C is the unsaturated fatty acid.

15. The type, number, and order of amino acids linked together

16. The nucleotide's base

17. Hydrogen bonding

SOLUTIONS TO THE TEST FOR MODULE #6

1. a. Cytoplasmic streaming - The motion of the cytoplasm which results in a coordinated movement of the cell's organelles

b. Isotonic solution - A solution in which the concentration of solutes is essentially equal to that of the cell which resides in the solution

c. Cytolysis - The rupturing of a cell due to excess internal pressure

d. Phagocytosis - The process by which a cell engulfs foreign substances or other cells

e. Activation energy - energy necessary to get a chemical reaction going

f. Plasmolysis - A collapse of the cell's cytoplasm due to lack of water

2. secretion

3. the middle lamella

4. Digestion breaks down big molecules for the purpose of respiration. Respiration breaks down little molecules for the purpose of producing energy.

5. There are 6 correct answers and 2 acceptable ones. The student needs only to identify 2: ribosomes, chloroplasts, smooth endoplasmic reticulum, rough endoplasmic reticulum, plastids, and Golgi bodies all have direct roles in biosynthesis. You can also count cytoplasm and nucleus, since they are involved in all cellular functions, including biosynthesis.

6. Rough ER has ribosomes, whereas smooth ER does not.

7. A chromoplast

8. The lysosome

9. Yes, but it will only do glycolysis

10. Aerobic respiration

11. The electron transport system

12. The fact that phospholipids have a hydrophilic end and a hydrophobic end (If the student mentions phosopholipids at all, count it right).

13. uracil, adenine, cytosine

14. 3

15. adenine, cytosine, guanine, guanine, cytosine, thymine, adenine, thymine, thymine

SOLUTIONS TO THE TEST FOR MODULE #7

1. a. Interphase - The time interval between cellular reproduction

b. Karyotype - The figure produced when the chromosomes of a species during metaphase are arranged according to size

c. Diploid cell - A cell whose chromosomes come in homologous pairs

d. Gametes - Haploid cells (n) produced by diploid cells (2n) for the purpose of reproduction

e. Virus - A non-cellular infectious agent that has two characteristics:
(1) It has genetic material inside a protective protein coat
(2) It cannot reproduce itself

2. Environmental factors and spiritual factors

3. The proteins help coil and support the DNA so that it stays wound up in the nucleus.

4. a. anaphase
b. prophase
c. telophase
d. metaphase

The proper order is prophase, metaphase, anaphase, telophase.

5. The diploid number is 24, the haploid number is 12

6. In mitosis, one cell becomes two cells identical to the original. Thus, if 3 cells undergo mitosis, 6 cells result. Since they are identical to the original cell, they will each have 32 chromosomes total

7. In meiosis, one diploid cell becomes 4 haploid cells. Thus 12 cells result, and they will have half of the original chromosome number. They will have 16 total chromosomes each.

8. Meiosis II, which is essentially mitosis performed on haploid cells.

9. Meiosis II starts with a haploid cell and pulls apart the duplicated chromosomes.

10. Mitosis starts with diploid cells and ends with diploid cells.

11. Only in female meiosis are three of the fours gametes useless.

12. Either of the following are acceptable: A virus cannot reproduce on its own; or A virus has no means of taking in nutrients and converting them into useable energy.

13. It stimulates the body into producing antibodies against the virus.

14. DNA and RNA

SOLUTIONS TO THE TEST FOR MODULE #8

1.

a. True breeding - If an organism has a certain characteristic that is always passed on to all of its offspring, we say that this organism bred true with respect to that characteristic.

b. Allele - One of a pair of genes that occupies the same position on homologous chromosomes

c. Homozygous genotype - A genotype in which both alleles are identical

d. Heterozygous genotype - A genotype with two different alleles

e. Recessive allele - An allele that will not determine the phenotype unless the genotype is homozygous with that allele

f. Monohybrid cross - A cross between two individuals concentrating on only one definable trait

g. Dihybrid cross - A cross between two individuals concentrating on two definable traits

h. Autosomal inheritance - Inheritance of a genetic trait not on a sex chromosome

2.

1. The traits of an organism are determined by its genes.

2. Each organism has two alleles that make up the genotype of a given trait.

3. In sexual reproduction, each parent contributes ONLY ONE of its alleles to the offspring.

4. In each genotype, there is a dominant allele. If it exists in an organism, the phenotype is determined by that allele.

3. RR, Rr

4. One

5. Two

6. a. Tt

   b. tt

c.

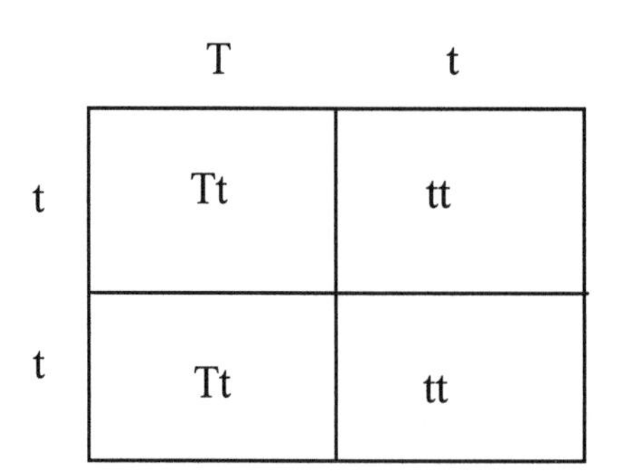

(which genotype goes on top is irrelevant)

7. Filled circles and squares represent those who cannot roll their tongues. Remember, if the same trait is expressed in both parents and they give rise to the other trait, then they must each carry a recessive allele. Look at the first set of parents. They are both represented by empty shapes. Yet, they give rise to a child with a filled square. Thus, the empty shapes must represent the dominant allele and the filled must represent the recessive.

8. a. *YT, Yt, yT, yt*

b. (The order of gametes is irrelevant)

	YT	*Yt*	*yT*	yt
YT	*YYTT*	*YYTt*	*YyTT*	*YyTt*
Yt	*YYTt*	*YYtt*	*YyTt*	*Yytt*
yT	*YyTT*	*YyTt*	*yyTT*	*yyTt*
yt	*YyTt*	*Yytt*	*yyTt*	*yytt*

c. tall, yellow peas (genotypes YYTT, YyTt, YyTT, YYTt) 9 of 16 or 56.25 %
tall, green peas (genotypes yyTT, yyTt) 3 of 16 or 18.75 %
short, yellow peas (genotypes YYtt, Yytt) 3 of 16 or 18.75 %
short, green peas (genotype yytt) 1 of 16 or 6.25 %

9. In sex-linked traits, there is no allele on the Y chromosome. Thus, males have only one allele. Your chance of expressing the recessive phenotype is better if you have only one allele.

10. a. 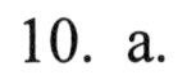

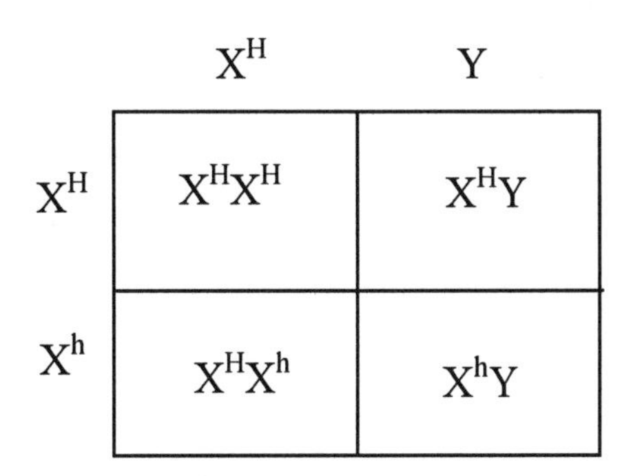

b. 0%

c. 50%

11. Genetics is only one part of a person's makeup. Environmental and spiritual factors play a role as well.

12. The student needs 4 of the following: Autosomal inheritance, sex-linked inheritance, change in chromosome structure, mutation, or change in chromosome number.

SOLUTIONS TO THE TEST FOR MODULE #9

1.

a. The immutability of species - The idea that each individual species on the planet was specially created by God and could never fundamentally change

b. Microevolution - The theory that natural selection can, over time, take an organism and transform it into a more specialized species of that organism.

c. Macroevolution - The hypothesis that the same processes which work in microevolution can, over eons of time, transform an organism into a completely different kind of organism

d. Strata - Distinct layers of rock

e. Fossils - Preserved remains of once-living organisms

f. Structural Homology - The study of similar structures (bones or organs, for example) in different species

2. The *HMS Beagle*

3. The concept of a struggle for survival and the present is the key to the past

4. Microevolution: The horses remained horses, they just got faster.

5. Macroevolution: The bacteria evolved into a completely different organism.

6. None

7. Structural homology, the fossil record, and molecular biology

8. Darwin proposed the theory of microevolution in that book. It is a well-documented scientific theory today. The part of the book that was wrong was his macroevolutionary ideas.

9. The ape's should be closer to the human's, because according to the wild idea of macroevolution, the ape is closer in lineage to the man than is the rat.

10. Neo-Darwinism uses mutations to add information to the genetic code.

11. Punctuated equilibrium attempts to explain why there are no intermediate links in the fossil record.

12. The sequence in (a) is closest to the sequence of interest.

SOLUTIONS TO THE TEST FOR MODULE #10

1.
a. Ecosystem - An association of living organisms and their physical environment

b. Biomass - A measure of the mass of organisms within a region divided by the area of that region

c. Watershed - An ecosystem where all water runoff drains into a single river or stream

d. Transpiration - Evaporation of water from the leaves of a plant

e. Greenhouse effect - The process by which certain gases (principally water, carbon dioxide, and methane) trap heat that would otherwise escape the earth and radiate into space

2. The frog population would grow out of control because they would have no predators. NOTE: Give partial credit if the student simply says that the ecosystem will be thrown out of balance.

3. Secondary consumer, tertiary consumer

4. Secondary consumer, tertiary consumer

5. The key here is that each level should be half as wide as the one under it.

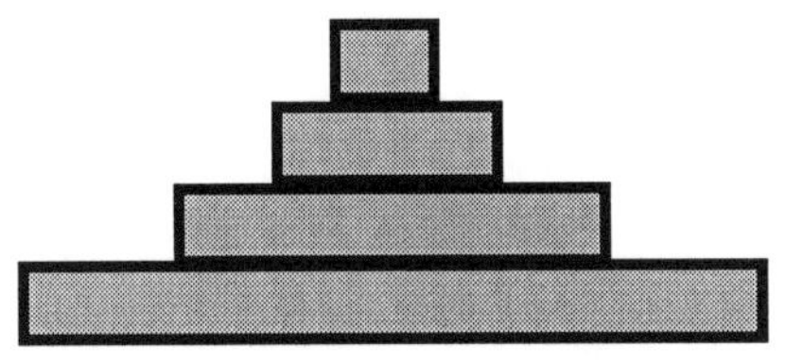

6. The student need only give two of these three:

The clownfish and the sea anemone form a symbiotic relationship. The clownfish is protected by the sea anemone and it attracts food to the sea anemone.

The goby and the blind shrimp have a symbiotic relationship in which the goby protects the blind shrimp, and the blind shrimp provides a home for the goby.

The Oriental sweetlips and blue-streak wrasse form a symbiotic relationship in which the sweetlips gets it teeth cleaned by the wrasse and the wrasse gets food from the sweetlips' teeth.

7. Too many nutrients in the water indicates that there are no plants to regulate the nutrient flow in the water cycle. The most likely explanation is deforestation.

8. It adds oxygen to the air.

9. Carbon dioxide can be dissolved in the ocean.

10. Without surface runoff, more water would evaporate from the ocean than what gets returned to it. The ocean would begin losing water.

11. The greenhouse effect is a good thing because without it, the earth would be too cold for life to survive. Global warming is an enhancement of the greenhouse effect that causes the earth to get too warm. Thus, it really is too much of a good thing.

12. No

SOLUTIONS TO THE TEST FOR MODULE #11

1. a. Invertebrates -Animals that lack a backbone

b. Vertebrates - Animals that possess a backbone

c. Nematocysts - Small capsules that contain a toxin which is injected into prey or predators

d. Posterior end - The end of an animal that contains the tail

e. Hermaphroditic - Possessing both the male and the female reproductive organs

f. Mantle - A sheath of tissue that encloses the vital organs of a mollusk, secretes its shell, and performs respiration

2. a. Porifera b. Mollusca c. Cnidaria d. Annelida, e. Platyhelminthes

3. This is a bivalve, because it has two shells.

4. Sponges eat the algae, bacteria, and organic matter that is in the water which they pump through their bodies. The student need only list one of the food types, but he or she must say where they come from.

5. Hard, prickly sponges contain spicules while soft sponges contain spongin.

6. The organisms of phylum Cnidaria have nematocysts.

7. a. ganglia b. esophagus c. oviduct d. dorsal blood vessel e. clitellum
f. seminal vesicle g. ventral nerve chord

8. It has not mated yet. When an earthworm mates, its seminal vesicles are emptied and its seminal receptacles get filled.

9. The cuticle allows the earthworm to get oxygen and release carbon dioxide. The student could just say that it performs respiration.

10. The second planarian is the parasite. Parasites can have simple nervous systems because they need not search for food.

11. Members of phylum Cnidaria asexually reproduce by budding.

12. Members of phylum Platyhelminthes asexually reproduce by regeneration.

SOLUTIONS TO THE TEST FOR MODULE #12

1.

a. Exoskeleton - A body covering, typically made of chitin, that provides support and protection

b. Simple eye - An eye with only one lens

c. Open circulatory system - A circulatory system that allows the blood to flow out of the blood vessels and into various body cavities so that the cells are in direct contact with the blood

d. Statocyst - The organ of balance in a crustacean

e. Gonad - A general term for the organ that produces gametes

2. a. antennae b. carapace c. swimmerets

3. a. stomach b. heart c. pericardial sinus d. anus e. nerve chord f. digestive gland g. green gland

4. Arthropods molt because their growing bodies get too large for their exoskeleton.

5. Four pairs of walking legs, two segments in body, no antennae, book lungs, four pairs of simple eyes.

6. Three pairs of walking (cr jumping) legs, wings, three segments in body, one pair of antennae.

7. egg, larva, pupa, adult

8. A complex network of tracheae connected to spiracles in the exoskeleton takes the place of a respiratory system in insects.

9. Coleoptera

10. Orthoptera

SOLUTIONS TO THE TEST FOR MODULE #13

1. a. Bone marrow - A soft tissue inside the bone that produces blood cells

b. Appendicular skeleton - The portion of the skeleton that attaches to the axial skeleton and has the limbs attached to it

c. Cerebrum - The lobes of the brain that integrate sensory information and coordinate the creature's response to that information

d. Cerebellum - The lobe that controls involuntary actions and refines muscle movement

e. Anadromous - A lifecycle in which creatures are hatched in fresh water, migrate to salt water as adults, and then go back to fresh water in order to reproduce

f. Bile - A mixture of salts and phospholipids that aids in the breakdown of fat

g. Ectothermic - Lacking an internal mechanism for regulating body heat

h. Hibernation - A state of extremely low metabolism

2. There are many possible answers here. Lampreys, Sea squirts, frogs, toads, and salamanders are possible answers. The student can have any two.

3. a. Class Chondrichthyes b. Class Osteichthyes c. Subphylum Urochordata d. Class Amphibia e. Class Amphibia f. Subphylum Cephalochordata g. Class Agnatha

4. Sight is its weakest sense, because the optic lobes control sight.

5. Hemoglobin

6. Fertilization is external and the development is oviparous.

7. The shark has a cartilaginous skeleton. The other two have bony skeletons. Thus, the shark has the most flexible skeleton.

8. a. Spinal cord b. Air bladder c. Gonad d. Liver

9.

Organ	Basic Function
Spinal cord	Sends messages from brain to other parts of the body and vice-versa
Air bladder	Allows fish to change depths and float in water
Gonad	Reproduction
Liver	Makes bile for the digestion of fats

10. a. artery b. vein c. artery

11. a. away from heart b. to heart c. away from heart

12. a. oxygen rich b. oxygen poor c. oxygen poor

13. It is a frog.

14. An amphibian's skin is its most important respiratory organ.

SOLUTIONS TO THE TEST FOR MODULE #14

1. a. Vegetative organs - The stems, roots, and leaves of a plant

b. Reproductive plant organs - The flowers, fruits, and seeds of a plant

c. Undifferentiated cells - Cells that have not specialized in any particular function

d. Xylem - A vascular tissue that carries water and dissolved substances upward in a plant

e. Phloem - Vascular tissue that carries water and dissolved substances downward in a plant

f. Leaf mosaic - The arrangement of leaves on the stem of a plant

g. Leaf margin - The characteristics of the leaf edge

h. Deciduous plant - A plant that loses its leaves before winter

2.

Letter	Shape	Margin	Venation
a.	Deltoid	Serrate	Pinnate
b.	Chordate	Entire	Pinnate
c.	Linear	Entire (count Serrate as right also, because the edge is jagged.)	Parallel
d.	Lobed	Dentate	Palmate
e.	Circular	Undulate	Pinnate

3. They open and close the stomata.

4. The top layer will be lighter. The side of the leaf with the spongy mesophyll is lighter than the side with the palisade mesophyll.

5. The abscission layer.

6. Anthocyanin is a pigment that can give a leaf a color other than green.

7. The most growth in a root takes place in the meristematic tissue. Remember, meristematic tissue can perform mitosis, which is necessary for growth.

8. It is from a monocot. The face-like shape indicates this.

9. Bark is cracked because the stem grows and breaks the bark when it gets too big.

10. Trees that produce cones are members of phylum Coniferophyta.

11. This plant has a taproot system.

12. The number of cotyledons produced in the seed is the fundamental difference between monocots and dicots. In monocots, the venation is parallel while it is netted in dicots. The fibrovascular bundles are packaged differently in monocots and dicots. The root and stem structures are different. Typically, monocots have fibrous root systems whereas dicots have taproot systems. Finally, monocots usually produce flowers in groups of 3 or 6 while dicots produce flowers in groups of 4 or 5. The student need list only two of these.

SOLUTIONS TO THE TEST FOR MODULE #15

1. a. Physiology - The study of life processes that occur in the daily life of an organism

b. Nastic movement - Movement in a plant caused by changes in turgor pressure

c. Pore spaces - Spaces in the soil which determine how much water and air the soil contains

d. Loam - A mixture of gravel, sand, silt, and clay

e. Gravotropism - A growth response to gravity

f. Imperfect flowers - Flowers with either stamens or carpels, but not both

2. A plant uses water for photosynthesis, turgor pressure, hydrolysis, and transport.

3. No. The flow of water and minerals through the xylem is most likely caused by the evaporation of water through the stomata, as dictated by the cohesion-tension transport theory.

4. Yes. The flow of materials through the phloem is controlled by phloem cells and is unrelated to the position of the stomata. The flow will *eventually* stop, because the plant will starve due to a lack of photosynthesis. That will take a long while, because plants store up excess food.

5. It came from the phloem. The xylem transport mostly water and minerals, whereas the phloem transport organic substances.

6. No. Insectivorous plants do not use the insects they catch for food. They use them for the raw materials of biosynthesis.

7. The parent reproduced vegetatively. Vegetative reproduction is asexual, which results in a genetic copy.

8. a. Ovule b. Anther c. Sepal d. Petal

9. b

10. c

11. d

12. a

13. Cotyledons perform photosynthesis after germination. These are the first photosynthetic leaves of a plant.

14. The fruit allows for the dispersal of seeds to places away from the parent.

15. There are many possible answers. The student needs at least 3:

wind, bees, beetles, birds, moths, or butterflies

16. An embryo sac has more cells. A pollen grain usually has 3: two sperm cells and a tube nucleus. An embryo sac has 7: 6 haploid eggs and a double-nucleus cell that becomes the endosperm.

17. The zygote is diploid, because it is the result of a union between two haploid cells. The endosperm is the result of a union between 3 haploid nuclei, so it is a "3n" cell.

SOLUTIONS TO THE TEST FOR MODULE #16

1.

a. Amniotic egg - An egg in which the embryo is protected by a membrane called an amnion. In addition, the egg is covered in a hard or leathery covering.

b. Hemotoxin - A poison that attacks the red blood cells and blood vessels, destroying circulation

c. Endothermic - A creature is endothermic if it has an internal mechanism by which it can regulate its own body temperature, keeping it constant.

d. Placenta - A structure that allows nutrients and gases to pass between the mother and the embryo

e. Gestation - The period of time during which an embryo develops before being born

2. It is a bird. Not all bird characteristics are listed, but no vertebrates other than birds have all of the ones listed in this question.

3. It is a reptile. Not all reptile characteristics are listed, but no vertebrates other than reptiles have all of the ones listed in this question.

4. It is a mammal. Not all mammal characteristics are listed, but no vertebrates other than mammals have all of the ones listed in this question.

5. a. Rhynchocephalia b. Squamata c. Crocodilia d. Testudines

6. No. Reptiles are ectothermic.

7. Contour feathers have hooked barbules.

8. You have found a reptile egg. Bird eggs are covered with a hard shell.

9. The bird should start preening. This will oil the feathers, making the hooked barbules slide easily on the smooth barbules.

10. The underhair will be thicker, because the main function of underhair is insulation.

11. The first species has the shorter gestation period. The longer the gestation period, the more developed the offspring are at birth. Since the second species gave birth to well-developed offspring, it must have had a long gestation period.

12. a. amnion b. yolk sac c. allantois